S0-DUV-653

SCANNING ELECTRON MICROSCOPY

THE SYSTEMATICS ASSOCIATION PUBLICATIONS

1. BIBLIOGRAPHY OF KEY WORKS FOR THE IDENTIFICATION OF THE BRITISH FAUNA AND FLORA
3rd edition (1967)
Edited by G. J. KERRICH, R. D. MEIKLE *and* NORMAN TEBBLE

2. THE SPECIES CONCEPT IN PALAEONTOLOGY (1965)
Edited by P. C. SYLVESTER-BRADLEY, B.Sc., F.G.S.

3. FUNCTION AND TAXONOMIC IMPORTANCE (1959)
Edited by A. J. CAIN, M.A., D.Phil., F.L.S.

4. TAXONOMY AND GEOGRAPHY (1962)
Edited by DAVID NICHOLS, M.A., D.Phil.

5. SPECIATION IN THE SEA (1963)
Edited by J. P. HARDING *and* NORMAN TEBBLE

6. PHENETIC AND PHYLOGENETIC CLASSIFICATION (1964)
Edited by V. H. HEYWOOD, Ph.D., D.Sc. *and* J. McNEILL, B.Sc., Ph.D.

7. ASPECTS OF TETHYAN BIOGEOGRAPHY (1967)
Edited by C. G. ADAMS *and* D. V. AGER

8. THE SOIL ECOSYSTEM (1969) *Edited by* J. G. SHEALS

LONDON. Published by the Association

Price per volume: No. 1, 30s (cloth), 24s (paper); Nos. 2–6, 20s; No. 7, 36s; No. 8, 50s. (U.S.A. No. 1, $4.25 (cloth), $3.50 (paper); Nos. 2–6, $3; No. 7, $5; No. 8, $6.60. Free of charges)

Available from E. W. Classey Ltd., 353 Hanworth Road, Hampton, Middlesex

SYSTEMATICS ASSOCIATION SPECIAL VOLUMES

1. THE NEW SYSTEMATICS (1940)
Edited by JULIAN HUXLEY (Reprint in preparation)

2. CHEMOTAXONOMY AND SEROTAXONOMY (1968)*
Edited by J. G. HAWKES

3. DATA PROCESSING IN BIOLOGY AND GEOLOGY (1971)*
Edited by J. L. CUTBILL

*Published by Academic Press for the Systematics Association

THE SYSTEMATICS ASSOCIATION
SPECIAL VOLUME No. 4

SCANNING ELECTRON MICROSCOPY

SYSTEMATIC AND EVOLUTIONARY APPLICATIONS

*Proceedings of an International Symposium
held at the Department of Botany, University of Reading*

Edited by

V. H. HEYWOOD

Department of Botany, University of Reading, England

1971

Published for the
SYSTEMATICS ASSOCIATION
by
ACADEMIC PRESS · LONDON · NEW YORK

ACADEMIC PRESS INC. (LONDON) LTD
Berkeley Square House
Berkeley Square,
London, W1X 6BA

U.S. Edition published by
ACADEMIC PRESS INC.
111 Fifth Avenue,
New York, New York 10003

Copyright © 1971 By ACADEMIC PRESS INC. (LONDON) LTD

All Rights Reserved

No part of this book may be reproduced in any form by photostat, microfilm, or any other means, without written permission from the publishers

Library of Congress Catalog Card Number: 76–149699

ISBN: 0 12 347050 1

PRINTED IN GREAT BRITAIN BY
W. S. COWELL LTD, BUTTER MARKET, IPSWICH

List of Contributors

ALVIN, K. L., *Department of Botany, Imperial College of Science and Technology, London, England* (p. 297)

BOULTER, M. C., *Department of Biological Science, North East London Polytechnic, London, England* (p. 211)

ECHLIN, PATRICK, *Botany School, University of Cambridge, Cambridge, England* (p. 307)

GRIFFITHS, D. A., *Pest Infestation Laboratory, Agricultural Research Council, Slough, Buckinghamshire, England* (p. 67)

HAWKER, LILIAN E., *Department of Botany, The University of Bristol, Bristol, England* (p. 237)

HAY, WILLIAM W., *Institute of Marine and Atmospheric Sciences, Miami, Florida, U.S.A.* (p. 123)

HEIM, ROGER, *Muséum National d'Histoire Naturelle, Paris, France* (p. 251)

HEYWOOD, V. H., *Department of Botany, University of Reading, Reading, England* (p. 1)

HINTON, H. E., *Department of Zoology, The University of Bristol, Bristol, England* (p. 17)

NOËL, DENISE, *Laboratoire de Géologie, Muséum National d'Histoire Naturelle, Paris, France* (p. 113)

PERREAU, JACQUELINE, *Laboratoire de Cryptogamie, Muséum National d'Histoire Naturelle, Paris, France* (p. 251)

RAMSAY, A. T. S., *School of Environmental Sciences, University of East Anglia, Norwich, East Anglia, England* (p. 179)

REYRE, Y., *Laboratoire de Géologie, Muséum National d'Histoire Naturelle, Paris, France* (p. 145)

ROSS, R., *Botany Department, British Museum (Natural History), London, England* (p. 155)

SHEALS, J. G., *British Museum (Natural History), London, England* (p. 67)

SIMS, PATRICIA A., *Botany Department, British Museum (Natural History), London, England* (p. 155)

SYLVESTER-BRADLEY, P. C., *Department of Geology, The University of Leicester, Leicester, England* (p. 95)

VELTKAMP, C. J., *Hartley Botanical Laboratories, University of Liverpool, Liverpool, England* (p. 285)

WILLIAMS, A., *Department of Geology, Queen's University of Belfast, Belfast, Northern Ireland* (p. 37)

WILLIAMS, S. T., *Hartley Botanical Laboratories, University of Liverpool, Liverpool, England* (p. 285)

Preface

This volume is based on the papers presented at an international symposium on Scanning Electron Microscopy held at the University of Reading on 7-9 April 1970 under the aegis of the Systematics Association.

Impressed by the impact which scanning electron microscopy was having on biological research, the Council of the Systematics Association decided that it would be both timely and appropriate to hold a meeting to discuss the systematic and evolutionary importance of the results being obtained by this new technique and to consider the ways in which it could be used and developed for future research.

Although scanning electron microscopes only became available commercially towards the end of 1965, biologists were quick to appreciate their value in studying surface details in a wide range of materials, and by 1969 an extensive series of papers describing the application of scanning electron microscopy to systematic and related research had been published. Although many of these papers were superficial and did little more than demonstrate the capability of the scanning microscope to produce beautiful and often dramatic, quasi-three dimensional micrographs illustrating surface features, sufficient detailed research in depth was produced by some workers to indicate the actual or potential value of the machine in providing important new information or in leading to reinterpretations of many structures that had hitherto been observed only vaguely by conventional light microscopy. In addition it soon became evident that the sheer increase in the information produced in scanning micrographs had serious implications for descriptive taxonomy and the illustration of taxonomic papers in many groups.

In planning the symposium the aim was to give examples of as many groups as possible, plant and animal, extant and fossil, where scanning electron microscopy had led to a considerable advance in our understanding of their structure, systematics and evolution. Unfortunately it was not possible to cover all fields but the range of papers included in this volume should allow a reasonable assessment of the role that scanning electron microscopy is coming to occupy in systematic and evolutionary biology. Two of the contributors—P. C. Sylvester-Bradley (Chapter 5) and W. W. Hay (Chapter 7) consider the information and publication problems referred to above. The last chapter, by P. Echlin, discusses preparation techniques for labile material which is a major technical problem. Other contributors also refer to problems and techniques of specimen preparation.

The symposium was attended by 120 participants from eight countries. In the organization of the meeting I would like to record my gratitude to Dr. D. M.

Moore and Dr. Barbara Pickersgill for their invaluable assistance. Thanks are due also to the University of Reading for its hospitality and for the reception held for the participants on Wednesday 8 April.

The task of editing this volume for the press has been greatly lightened by the expert guidance of the production department of Academic Press while the high quality work of the printers who have produced such a handsome volume is available for all to see.

V. H. Heywood

Department of Botany
University of Reading
December, 1970

Contents

	PAGE
THE SYSTEMATICS ASSOCIATION PUBLICATIONS	ii
LIST OF CONTRIBUTORS	v
PREFACE	vii

1 The Characteristics of the Scanning Electron
  Microscope and Their Importance in Biological Studies 1
  V. H. HEYWOOD

2 Polyphyletic Evolution of Respiratory
  Systems of Eggshells, with a Discussion of
  Structure and Density-independent and
  Density-dependent Selective Pressures 17
  H. E. HINTON

3 Scanning Electron Microscopy of the
  Calcareous Skeleton of Fossil and Living Brachiopoda 37
  ALWYN WILLIAMS

4 The Scanning Electron Microscope in
  Acarine Systematics 67
  D. A. GRIFFITHS and J. G. SHEALS

5 The Reaction of Systematics to the Revolution
  in Micropalaeontology 95
  P. C. SYLVESTER-BRADLEY

6 Intérêt du Microscope Électronique à
  Balayage dans la Définition des Critères
  Génériques chez les Coccolithophoridées Fossiles 113
  DENISE NOEL

7 Scanning Electron Microscopy and Information
  Transfer in Systematic Micropaleontology 123
  WILLIAM W. HAY

8   Interprétation Botanique des Pollens
    Inaperturés du Mésozoïque Saharien.
    Essai de Classification d'après l'Observation
    en Microscopie Électronique à Balayage   145
                Y. REYRE

9   Generic Limits in the Biddulphiaceae
    As Indicated by the Scanning Electron Microscope   155
                R. ROSS and PATRICIA A. SIMS

10 The Study of Lower Tertiary Calcareous
    Nannoplankton from the North Atlantic
    Ocean by Means of Scanning Electron Microscopy   179
                A. T. S. RAMSAY

11 Fine Details of Some Fossil and Recent Conifer Leaf Cuticles   ..   211
                M. C. BOULTER

12 Scanning Electron Microscopy of Fungi
    and its Bearing on Classification   237
                LILIAN E. HAWKER

13 Étude Ornementale de Basidiospores au
    Microscope Électronique à Balayage   251
                ROGER HEIM et JACQUELINE PERREAU

14 The Value of Scanning Electron Microscopy
    for the Examination of Actinomycetes   285
                S. T. WILLIAMS and C. J. VELTKAMP

15 The Study of Fossil Epiphyllous Fungi by
    Scanning Electron Microscopy   297
                K. L. ALVIN

16 Preparation of Labile Biological Material
    for Examination in the Scanning Electron Microscope   307
                PATRICK ECHLIN

AUTHOR INDEX   317
SUBJECT INDEX   323

1 | The Characteristics of the Scanning Electron Microscope and Their Importance in Biological Studies

V. H. HEYWOOD

*Department of Botany, The University of Reading,
Reading, England*

"The objective of natural science is to discover the relations of things and events at different levels of complexity."—C. F. A. Pantin, "The Relations between the Sciences", 1968.

Abstract: The method of image formation with the scanning electron microscope, giving easily comprehensible, quasi-three-dimensional representations of objects examined, at a wide range of magnifications, is probably its most valuable characteristic for biological research, especially in the fields of systematics, morphology and evolutionary interpretation. This leads to a better understanding of the spatial relations of features of microtopography, reveals unsuspected detail and previously undescribed characters. These points are illustrated with reference to fruit (mericarp) surfaces of the Umbelliferae-Caucalideae.

The increased information made available in the cathode ray tube is discussed in relation to recent developments in the nature, selection and handling of characters in numerical taxonomy and in terms of Gestalt formation and computer methods of information processing.

The concept of the microhabitat—phyllosphere, carposphere (Heywood) and *nano-faciès* (Noël, 1967) is discussed.

Problems of publishing scanning electron micrographs and their role in systematic research *vis à vis* descriptions, line drawings etc., are discussed.

INTRODUCTION

The introduction of the scanning electron microscope (SEM) is causing major reappraisals in many areas of biology. It is having repercussions not only in morphology and anatomy but in systematics, ecology and evolutionary studies. Developments have been so rapid and the published results so diverse that it is extremely difficult to assess what role it will come to occupy, although I am personally confident that the SEM will become a routine tool in biological research and, in many fields, of much more value than the transmission electron microscope.

The principles of scanning electron microscopy are now familiar to most biologists and I shall not consider them in detail. I shall, however, attempt to consider those characteristics of image formation that have made the SEM so valuable in biological (and earth) sciences and some of the problems that have been caused by the increased information obtained.

With modern optical microscopes details of the order of 50 nm can be resolved under certain conditions although 300 nm is more usual, and with a stereoscopic binocular, of course, much poorer resolution is obtained. The depth of focus at such magnifications is very restricted and in many biological problems it is this rather than the limits of magnification and resolution that have blocked progress. The transmission electron microscope (TEM), with its enormously high resolution, has very great inherent depth of focus but it is largely wasted because of the ultrathin specimens used except in the case of replicas.

The SEM uses the secondary electrons produced by the scanning beam to give a life-like, almost three-dimensional image of the surface of the specimen being scanned. The best resolution obtained with the commercially available models is about 15 nm, although the normal range varies from 50–20 nm according to the accelerating voltage used. Higher resolutions have been obtained in experimental machines but there is a limitation imposed in commercial machines by the hairpin tungsten filament normally used as an electron source. No doubt a new generation of high resolution machines will be marketed in the next few years and this will open up new areas of investigation at present prevented by the low resolutions obtainable. Some of the initial resistance of electron microscopists to SEM on the grounds of low resolution were understandable but ill-founded: the kinds of material, the magnification ranges and the nature of the information obtainable are so different in SEM and TEM that the two are not in competition and indeed often complement each other.

The magnification range of the SEM is extraordinarily wide: from $\times 10$ to $\times 100,000$ or more. It overlaps the hand lens, binocular, stereoscopic and compound microscope at one end and the medium range TEM at the other. Except in some classes of organisms or structures most of the SE micrographs that have been published have been in the $\times 50$ to 10,000 range. Certainly the SEM fills the magnification gap between the light microscope and the TEM but it is perhaps better to think of it as extending the range (and quality) of the former.

The advance represented by the SEM may be compared with that represented by high fidelity stereo over the phonograph. Both are larger than life, better in a distorted way than the real thing!

The very great depth of focus obtainable with the SEM combined with the method of image formation (which makes recessed areas appear dark and projecting areas cast shadows and thus allows the human eye to interpret and readily comprehend the images obtained) gives its most valuable characteristics. This contrasts with the difficulties of interpreting two-dimensional transmission micrographs. An important feature of the SEM, as pointed out by Oatley (1966) is that the image represents a magnified view of what would be seen by looking along the incident beam.

It is this enhanced visual effect that also creates some of the major problems posed by SEM. We are accustomed to all kinds of limitations on our sensory capacities. As Pantin (1968) says "In our perception of objects . . . our senses only cover a certain range; our immediate perceptions are limited by 'our sensory spectrum'." Not only are our sources of information restricted but, what is more relevant here, so is the magnitude of the objects we see. We cannot perceive objects or events which are too small or too large, too quick or too slow unless our attention is drawn to them by indirect means. We become aware of phenomena outside our sensory spectrum by their indirect effects, especially by instruments—microscopes, chromatographs etc. I have discussed on a previous occasion such problems which face the taxonomist in the machine age (Heywood, 1968b).

In the case of recognizable three-dimensional images we have become accustomed to the limitations imposed by the naked eye, the hand lens, and the stereoscopic binocular. Information or characters beyond the levels permitted by these is known to us in a highly distorted form, if at all, so that our knowledge of microtopology is very limited. Now the SEM has opened up new vistas to us and we are faced with the curious situation of "seeing" microfeatures in depth, in their correct spatial relationships for the first time and the only background we have for comprehending them comes from lower orders of magnification or from reconstructions. It is no wonder, therefore, that faced with an SE micrograph we are often at a loss to understand exactly what it is we see.

We have to learn to recognize objects at magnifications to which we are not accustomed. We find that apparently simple structures are extremely complex. We may find new kinds of microstructure not previously recognized. All these points can be illustrated by reference to work we have been carrying out for several years on the scanning electron microscopy of fruits of the Umbelliferae tribe Caucalideae (Heywood, 1968b, 1969) as part of a larger programme on the multivariate taxonomy of the group (McNeill *et al.*, 1969; Crowden *et al.*, 1969).

The complexity in microtopography of apparently simple structures is a characteristic feature revealed by scanning these fruits. This group of Umbelliferae is classified at generic level largely by the structure and morphology of the fruits, reliance being placed on such features as the number of primary and secondary ridges and the nature and distribution of the spines borne on them, as well as on internal features such as the curvature of the embryos, disposition and development of the secretory canals, vascular system, crystals etc.

From a gross morphological point of view we can talk of the group of *Daucus*-like Umbellifers with spiny fruits, much in the way that we talk about yellow-flowered Crucifers like *Brassica*. In practice it is only by looking carefully at the fruits that we can usually tell the genus to which any particular specimen belongs—by counting the number of ridges (primary and secondary) and looking at the number of rows of spines, and whether they are glochidiate at the apex or not. Conventional descriptions of the fruits are remarkably poor and have reduced even the complexity seen by the stereoscopic dissecting microscope to an unwarranted level of simplicity. In fact most of the species concerned were described when only primitive optical aids were available and features such as the primary ridges must have been observed originally from anatomical studies of transverse sections since they are far from obvious under the hand lens or low power microscope. In other words, taxonomists proceeded by a judicious mixture of extrapolation, imagination and pretence; indeed it was often the species that were recognized and the genus worked out from that! I doubt if many taxonomists have recognized a primary ridge in many of these fruits until shown them in scanning electron micrographs. This taxonomic process of reducing a complex pattern to a few simple characters is a matter I shall discuss in detail later.

Cerceau-Larrival (1962, 1965) looked at additional features in this group, such as cotyledon type, pollen type, seedling and various other vegetative characters. These, correlated with fruit structure, led her to propose a number of taxonomic rearrangements but the basic uncertainty as to the validity of many of the characters used remains. In our programme we have attempted to use a very wide range of features indeed from as many parts of the plant as possible, including an intensive survey of secondary plant constituents (phenolics, sugars, acetylenic compounds, terpenes, proteins etc.). Some of this work has been published but many of the data are still being studied.

When we came to the fruits (mericarps) we considered firstly the external features: apart from dimensions and the compression ratio (the latter a much confused character in the literature although a key feature in distinguishing, say, *Caucalis* from *Orlaya*), 12 features of the ridges and primary spines were recorded, 12 from the secondary spines, and a number of other features. Anatomical features added another 10–15 characters, giving a total of 40 or more for the fruits. All of these could be quite well observed by conventional light microscopy.

It was at this stage in 1966 that I became aware of the probable value of applying the new technique of SEM to these fruits with their remarkable topographic diversity and through the courtesy of the Cambridge Instruments Company I was able to examine a range of these Umbellifer fruits. The results, I recall, not only excited me but also the Cambridge operators of the instrument.

One of the immediate results of scanning the surface of the fruits was to reveal a wealth of detail of microstructural features which was at first somewhat overwhelming. In fact it was difficult to reconcile the images revealed, such was their richness, with the impressions we had obtained by conventional means. Basically the features found on the fruits by scanning are: (1) general surface characteristics such as the ridges, valleys, spines etc., which dramatically brought to life all sorts of spatial relationships not hitherto appreciated; (2) a wealth of microdetail in the form of protuberances on spines and other surfaces which had previously been regarded as basically simple, and these protuberances in turn bore ornamentations of second and third order; (3) a series of new microstructures such as hairs, microspines, etc., not previously recorded; (4) patterns of wax extrusions (5) a wealth of extraneous organisms and structures such as fungal hyphae, spores and reproductive structures, pollen grains, insect eggs etc.

Good examples of the overall differences in landscape shown by fruits of related genera are found in *Torilis*, *Caucalis* and *Turgenia*, between which there has often been confusion. Examination by SEM at low magnifications shows at a glance how widely different they are. In *Torilis* (Plate I, Figs 1,2) the strongly tuberculate secondary spines and the well marked primary ridges with appressed hairs are characteristic. These latter hairs are of an unusual kind, flattened and strongly tuberculate, not previously recorded. They appear to be characteristic of all species of *Torilis* so far studied. They can be detected by the stereoscopic microscope now we know of their existence but the fact is they were not recorded in the past. In *Caucalis* (Plate II, Fig. 1) the etuberculate but striate secondary spines and the sparse protuberances on the primary ridges form a stark contrast. *Turgenia* (Plate II, Fig. 2) on the other hand shows an almost nightmarish complexity, both of the secondary spines, primary ridges and vallecular areas, with a series of castellated spines at all stages of development and demonstrating a remarkable phyllotactic-like arrangement individually. It has to be remembered that all these spines are parts of single cells; they are not multicellular but are aspects of cuticular diversity.

Although the spines of *Torilis* are much more complex than previously described, an even more remarkable example is found in the monotypic genus

Chaetosciadium (Plate III, Figs 1, 2 and 3). This was described in the taxonomic literature as having bristly fruits—indeed they are so bristly that it is difficult to see the body of the fruits. The general effect of the bristle-like hairs is to make the genus instantly recognizable as no other genus is ornamented in this way. When examined by the SEM these bristles were shown to possess a very unusual structure. They are made up of a series of tube-like elements, these tubes becoming free at the apex at "nodes". Details will be published in a later paper. Although *Chaetosciadium* is often placed close to *Torilis* the peculiar structure of the spines would not suggest a close relationship. Yet in other respects, such as fruit anatomy, development, chromosome number, chemical constituents, *Chaetosciadium* is very similar to *Torilis* so that it appears that here we have a case where micromorphology as revealed by SEM is not a reliable guide to affinity. It has also raised the interesting problem of how the *Torilis* kind of secondary spine would evolve into the *Chaetosciadium* type. Perhaps they are not so basically dissimilar as they at present appear. Certainly the SEM will play an important role in investigating the nature and development of spines, hairs and other cuticular and epidermal features.

As a taxonomist used to employing daily indumentum features I have for a long time felt unhappy about the level of precision with which different kinds of covering are grouped under the same terms, e.g. tomentose, hirsute, velutinous, lanate etc. The difficulty is that there are two aspects of indumentum which are often confused—the overall appearance and texture which determines the descriptive term used and the nature of the individual hairs which might be quite diverse in different examples but give, in mass, the same overall effect. As a general rule, the denser the hairs, the more attention is paid to the indumentum as a whole, and the less to the type of individual hairs. An exception is when some or all of the hairs are gland-tipped. In many groups, the nature of the indumentum (as opposed to hairs) is used as a taxonomic, often diagnostic, feature, and I suspect that we have been misled by oversimplified descriptions. When detailed microscopical and developmental studies have been made of hairs and scales, valuable features have been detected (Plate IV, Figs 1, 2) and I

Plate I

Fig. 1. General view of part of mericarp surface of *Torilis* sp. showing primary ridges bearing tuberculate hairs, alternating with the large tuberculate bases of spines on the secondary ridges.
× 240.

Fig. 2. *Torilis* sp. Detail of base of tuberculate hair on primary ridge.
× 1700.

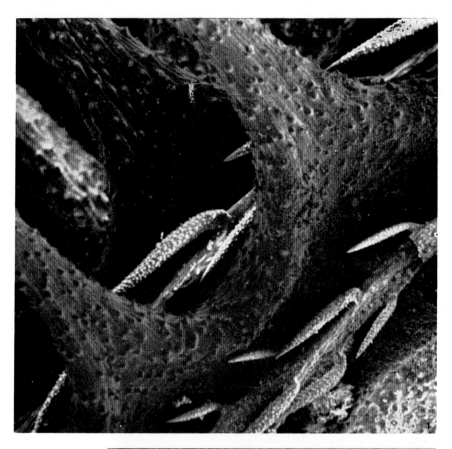

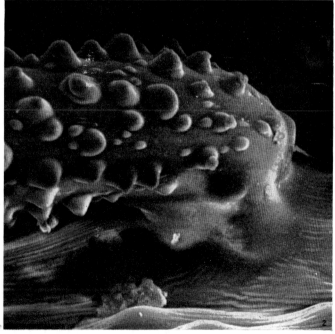

believe that SEM will have an important role to play here. Not only will the hairs be better observed but they will not be subjected to distortion and deformation as a result of their preparation for microscopic study at high levels of magnification.

HANDLING THE NEW INFORMATION

Although the information obtained on the cathode ray tube or micrograph through SEM can be used at the somewhat superficial level just described to assist in the solution of taxonomic problems, by confirming, modifying or rejecting previous decisions arrived at on general morphological or other grounds, detailed analysis and interpretation presents many problems. For one thing, we do not have a suitable terminology for many of the structures found, although attempts have been made for some features, e.g. wax types by Amelunxen *et al.* (1969), pollen ektexine patterns (Burrichter *et al.*, 1968; Reyre, 1968). How we should proceed raises much more fundamental issues.

We have to accept that, with the great increase in the amount of information displayed on the cathode ray tube over that obtainable by conventional optical microscopy, we have, in several senses, an information problem.

How we handle this increased information depends to a large extent, of course, on the purpose of our investigation. It also relates to many of the problems regarding the nature, handling and selection of taxonomic characters, the psychology of perception, classificatory procedures, information processing and computer methods, which have become central issues in the relatively new field of numerical taxonomy. It is fortunate that the information problem in SEM should come to light at a time when research in these fields is so active.

We can attempt to handle the SEM information as conventional characters or character-complexes or simply as "pure" information without being broken down or interpreted as individual characters, using computer processing. But before discussing these alternatives further it would be profitable to consider for a moment the ways in which taxonomists normally handle characters.

Plate II

Fig. 1. General view of part of mericarp surface of *Caucalis platycarpos* L. showing a primary ridge in the foreground bearing striate hairs and the lower parts of the striate spines on the secondary ridges.
×40.

Fig. 2. General view of part of mericarp surface of *Turgenia latifolia* (L.) Hoffm. showing ridged, tuberculate spines on secondary ridges and a rich ornamentation of spine-masses between the ridges.
×80.

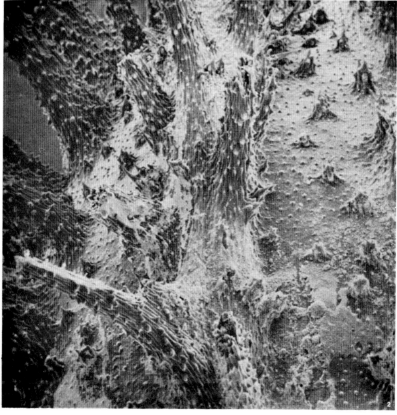

LIBRARY
LOS ANGELES COUNTY MUSEUM OF NATURAL HISTORY

TAXONOMIC CHARACTERS

As has frequently been pointed out, all taxonomic methods so far devised depend on some form of sampling. Since taxonomic situations are seldom defined with any precision and since organisms contain too much variation to be useful in many contexts, we resort to sampling to avoid this since complete descriptions would be too complex for the human mind to comprehend, even assuming that they could be achieved. We assume that it is possible to summarize the whole plant or animal, or particular parts, organs or organ systems in terms of characters (Watson *et al.*, 1966). We therefore select characters from those that are available and our descriptions are a compromise—they are abstracts or summaries for practical convenience (Davis and Heywood, 1963) that allow us to produce a level of grouping which permits information to be communicated although we seldom state exactly what use is to be made of these groupings—they are the intrinsic or general purpose classifications that we produce in the hope that they will be useful in a wide variety of contexts.

Characters have never been satisfactorily defined. This is not surprising when we consider that the living organism is a beautifully integrated unit and there is no reason why we should expect it for our convenience to lend itself for subdivision into what we call characters. One could go so far as to say that characters were "invented" by the taxonomist—they stem from our need to communicate (Cullen, 1968). This is readily understandable if we consider the traditional methods employed in taxonomy. Classification depends to a large extent on the Gestalt approach—groups are recognized according to the taxonomist's instinctive or neural impression of overall similarity. This is what has been termed the taxonomy of unanalysed entities: the human visual apparatus appreciates this Gestalt in a way we do not fully understand—whether it be at-a-glance appreciation of whole specimens, organ systems or individual organs. The complex mass of information reaching the brain is not broken down into

---

Plate III

Fig. 1. General view of whole mericarp of *Chaetosciadium trichospermum* (L.) Boiss. showing the dense cover of bristles.
$\times 30$.

Fig. 2. Tuberculate hairs on primary ridges of *Chaetosciadium*. Compare with Plate I, Fig. 1.
$\times 240$.

Fig. 3. Montage of a single spine (bristle) on the secondary ridges of *Chaetosciadium*, showing the tubular elements of which it is composed.
$\times 450$.

1. Characteristics of the SEM

characters, at least consciously, although some conspicuous features may be appreciated and may affect our decisions. Characters as such are not involved until it is necessary to communicate, describe and key—in other words to say precisely how the groups we have produced by imprecise means differ from one another. When we attempt to communicate our visual impressions we run into difficulties because not all characters are obvious—what we choose to regard as separate characters is not a fixed repertoire but varies according to the observer—we all have different pattern-elucidating capacities, hence the "born" taxonomist who is skilled at taxonomy without prior declaration of the classificatory procedure or ethic in use.

I have gone into these fundamental taxonomic procedures in some detail because they are highly relevant to our use and interpretation of information-rich scanning electron micrographs. Given two or more SE micrographs of fruits, seeds, pollen etc., we again use the Gestalt approach—grouping and separation can be made at a glance, as in the case of *Torilis* and *Caucalis* mericarp landscapes mentioned above. Whether or not we want to describe the variation we see, and if so to what extent, is a matter to which we have to give some considerable thought. In the case of pollen grains there is a tradition of detailed description of the pores, furrows etc., following the work of Erdtman, Faegri and others, and ektexine patterns have been described in detail by scanning electron microscopists, using complicated terminologies. One of the reasons for this is the need for detailed information in quaternary studies where the grains alone have to serve as a means of identification. I doubt if this is the only reason—I suspect another reason is simply that the variation in ektexine patterns is interesting and forms the end-point of important developmental studies in wall formation in pollen grains. Similar considerations may well apply to seeds and small fruits where the amount of surface variation is relatively small. Seed identification is economically important (e.g. weed seeds) as well as being necessary in Quaternary studies. In our Umbellifer fruits we have not attempted detailed description and only major differential characters have so far

---

Plate IV

Fig. 1. *Echium callithyrsum* Webb ex Bolle (Boraginaceae)—leaf indumentum of tuberculate hairs with swollen, compound, plate-like bases.
$\times 300$.

Fig. 2. *Sideritis canariensis* L. (Labiatae)—leaf indumentum showing compound, much branched hairs, some of the branches being gland-tipped. The indumentum is conventionally described as tomentose.
$\times 60$.

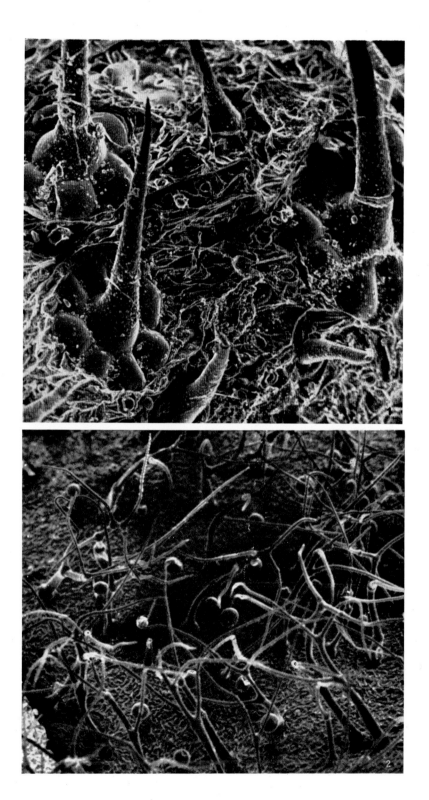

been singled out. The micrographs could possibly be used as an alternative to detailed descriptions but this would be a procedure without precedent in taxonomy, for the taxonomist will nearly always want to understand and analyse variation. On the other hand, with so many tasks to be completed, the micrographs could be used as a means of recording variation in a form which would be analysed later when time permits. This is an intriguing point which merits further discussion. Certainly the micrographs can be used as an alternative to illustrations such as line drawings, and should whenever possible be used to supplement descriptions.

If an attempt is made to get away from the Gestalt approach, which may break down in the case of closely similar organisms etc., characters have to be selected and described in detail: "unit characters" in the sense of numerical taxonomy. This is difficult enough with many of the normally available characters and we might be forgiven if we shirked the issue with scanning electron micrographs. There are alternative methods, however, whereby one could use the information on the cathode ray tubes simply as information—as patterns of so many dots—by computer processing without actually converting it into conventional recognizable characters. Research in this field is still in its infancy but the possibilities are wide—automatic sensing and data-recording devices have gradually been introduced in the past few years. Optical scanners, which digitize drawings, photographs, microscope slide preparations etc., could be applied to the scanning micrographs and the mass of information so derived computer processed and sorted into classifications, since it is indigestible by the traditional means we employ. The recent experiments of Sokal and Rohlf on the imaginary animals called Caminalcules so called after their inventor J. H. Camin are relevant. Agreement was recorded in visible structures over randomly selected minute areas of the images of pairs of organisms by making random masks from 25 punch cards each perforated with 25 randomly chosen holes and placing them over the black-and-white images of 2 groups of "organisms": (a) 29 recent species of Caminalcules (b) published illustrations of the pupae of 32 species of mosquitoes. Each illustration was overlaid with all the masks and each of the 625 holes was scored 1 when a black line appeared through it and 0 when no black showed. Illustrations were compared on the basis of matching scores for corresponding masks and holes and a numerical classification of the images showed surprising similarity to similar studies by conventional taxonomy or by numerical taxonomy based on the detailed description of characters. With considerable refinements similar techniques may prove valuable and acceptable for SEM images.

MICROHABITATS

One of the most exciting possibilities opened up by scanning electron microscopy is the study of microhabitats. Microbiologists have long investigated the so-called rhizosphere and more recently with the introduction of SEM, the phyllosphere—the detailed microhabitat provided by the leaf surface. During my studies on Umbellifer fruits I have been impressed by the rich and varied topography which forms a landscape habitat for various groups of organisms, particularly fungi, and I proposed the term *carphosphere* for this (Heywood, 1969). It is here that we have, I think, a clue to understanding the detailed microcharacters shown by the fruits in relation to adaptive value and natural selection. It has long been one of the pleasantries of fungal taxonomy that spore ornamentations are simply there to help the taxonomist. Not long ago it was generally stated that secondary plant constituents had no selective value but were simply by-products of the metabolic pathways and allowed plants to show off their diversity. Likewise details of fruit surfaces were normally regarded as of no functional significance. The trouble is, I suspect, that we were dealing with objects on a scale too small to visualize in the round. We still do not know the precise role of the fungi inhabiting these fruits—some may be there by accident but I am sure that some of them play a key role in fruit breakdown and decay which is, of course, of considerable selective value.

As many contributors to this symposium have discussed, we are seeking improved means of cleaning specimens for examination by SEM but it is by so doing that in some cases we are destroying one of the most interesting aspects that the specimens afford. Noël (Chapter 6) has found a similar problem with her work on the study of coccoliths *in situ* in rocks and has introduced the idea of the *nanofaciès*. It is in no way an exaggeration to say that SEM is opening up new horizons for us, albeit on a very micro scale, but none the less exciting or challenging for that.

ACKNOWLEDGEMENTS

All the scanning electron micrographs were made in the Electron Microscopy Laboratory, Department of Botany, University of Reading, on a JSM-2 instrument. The skilled technical assistance of Mr. S. K. Irtiza-Ali is gratefully acknowledged. I wish to thank Mr. D. Bramwell for providing the material on which Plate IV is based, and Dr. D. M. Moore for reading this paper in manuscript and proof.

REFERENCES

AMELUNXEN, F., MORGEN ROTH, K. and PICKSAK, T. (1969). Untersuchungen am der Epidermis mit den Stereoscan-Elektronmikroskop. *Z. Pflanzenphysiol.* **57**, 79–95.

BURRICHTER, E., AMELUNXEN, F., VAHL, J. and GIELE, T. (1968). Pollen- und Sporenuntersuchungen mit dem Oberflächen- Rasterelektronenmikroskop. *Z. Pflanzenphysiol.* **59**, 226–237.

CERCEAU-LARRIVAL, M.-T. (1962). Plantules et pollens d'Ombellifères. *Mem. Mus. Nat. Hist. Nat., Paris, ser. B (Bot.)* **14**, 1–166.

CERCEAU-LARRIVAL, M.-T. (1965). Le pollen d'Ombellifères Mediterranéennes III – Scandicineae Drude, IV-Dauceae Drude. *Pollen et Spores*, **7**, 35–62.

CROWDEN, R. K., HARBORNE, J. B. and HEYWOOD, V. H. (1969). Chemosystematics of the Umbelliferae—A general survey. *Phytochemistry* **8**, 1963–1984.

CULLEN, J. (1968). Botanical Problems of Numerical Taxonomy. *In* "Modern Methods in Plant Taxonomy" (Heywood, V. H., ed.), pp. 175–183, Academic Press, London and New York.

DAVIS, P. H. and HEYWOOD, V. H. (1963). "Principles of Angiosperm Taxonomy." Oliver & Boyd, Edinburgh and London.

HEYWOOD, V. H. (1968a). Plant taxonomy today. *In* "Modern Methods in Plant Taxonomy" (Heywood, V. H., ed.), pp. 3–12, Academic Press, London and New York.

HEYWOOD, V. H. (1968b). Scanning electron microscopy and microcharacters in the fruits of the Umbelliferae-Caucalideae. *Proc. Linn. Soc. London*, **179**, 287–289.

HEYWOOD, V. H. (1969). Scanning electron microscopy in the study of plant materials. *Micron* **1**, 1–14.

MCNEILL, J., PARKER, P. F. and HEYWOOD, V. H. (1969). A taximetric approach to the classification of the spiny-fruited members (tribe Caucalideae) of the flowering plant family Umbelliferae. *In* "Numerical Taxonomy" (Cole, A. J., ed.), pp. 129–147, Academic Press, London and New York.

OATLEY, C. W. (1966). The scanning electron microscope. *Sci. Prog.* **54**, 483–495.

PANTIN, C. F. A. (1968). "The Relations between the Sciences." Cambridge University Press, Cambridge.

REYRE, Y. (1968). La sculpture de l'exine des pollens des Gymnospermes et des Chlamydospermes et son utilization dans l'identification des pollens fossiles. *Pollen et Spores* **10**, 197–220.

WATSON, L., WILLIAMS, W. T. and LANCE, G. N. (1966). Angiosperm taxonomy: a comparative study of some novel numerical techniques. *J. Linn. Soc. Bot.* **59**, 491–501.

2 | Polyphyletic Evolution of Respiratory Systems of Eggshells, with a Discussion of Structure and Density-independent and Density-dependent Selective Pressures

H. E. HINTON

Department of Zoology, University of Bristol, Bristol, England

INTRODUCTION

The scanning electron microscope has made it possible to study the geometry of insect structures, particularly hard structures, at a magnification and resolution not possible before. In recent years an extensive survey has been made of the respiratory systems of insect eggshells with the aid of the scanning electron microscope. The terrestrial eggs of perhaps the majority of insects have meshworks in the chorion or shell that hold a layer of gas. Many aquatic eggs also have such meshworks but the percentage of aquatic eggs with distinct respiratory systems of this kind seems to be smaller. The species that have meshworks that hold a layer of gas in the inner part of the chorion also have aeropyles or holes that extend between the gas space and the outer layers of the shell and so effect the continuity of the chorionic layer of gas and the ambient atmosphere. Most aeropyles are easily recognized with the light microscope because they are one or more microns wide. The interstices between the struts of the chorionic meshworks that hold the gas layer are usually, as might be expected, wider than the mean free path of the respiratory gases, which is 0.1 μm for oxygen at 23 C°.

A very large number of terrestrial eggs, and a smaller number of aquatic ones, have the kind of permanent or unshrinkable physical gill called a plastron. The term plastron has been restricted to describe a gill that consists of a gas film of constant volume and an extensive water-air interface. Such films are held in position by a system of hydrofuge structures, and they resist wetting at the hydrostatic pressures to which they are normally subjected in nature. An egg with a plastron can remain immersed indefinitely and obtain the oxygen it requires from the ambient water provided that the water is well-aerated. The selective advantage of the plastron method of respiration can only be understood in relation to an environment that is alternately dry and flooded. The plastron provides a relatively enormous water-air interface for the extraction of oxygen dissolved in the ambient water. This enormous surface for the extraction of oxygen is provided without necessarily involving any reduction in the impermeability of much of the surface of the shell. When the egg is dry, the interstices of the plastron network provide a direct route for the entry of atmospheric oxygen that does not necessarily involve the egg in water loss over a large area because the connection between the plastron and the tissues of the egg, or the pharate larva, may be very restricted. The capacity to avoid loss of water when the environment is dry is therefore not necessarily impaired by the provision of a plastron. For instance, in eggs with respiratory horns, such as those of many species of *Drosophila* (Fig. 1a, b and c) the plastron network of the surface of the horns provides a very large respiratory area when the egg is flooded, but when

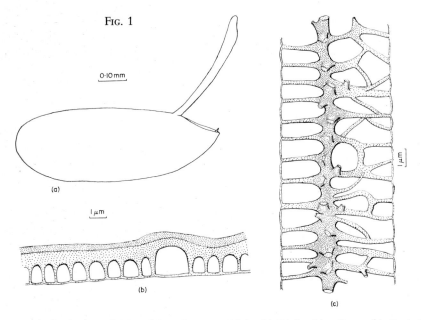

Fig. 1 (a, b, c). Egg of *Drosophila melanogaster* Meig. (a) Right side of egg. (b) Optical section of chorion showing the inner air spaces. (c) Optical section of dilated part of respiratory horn about 60 μm from its apex.

the egg is dry water is lost only through the cross sectional area at the base of the horn provided of course that the outer layer of the main body of the shell is impermeable. The general principles of plastron respiration have recently been discussed at length (Hinton, 1968, 1969) and do not require elaboration here.

Scanning electron microscope studies of the respiratory systems of a large number of eggs have revealed the fact that respiratory systems almost identical in fine structure have been independently evolved an enormous number of times. In this paper some of the principles of classification are considered in relation to polyphyletic origins. Because the functional significance of a high percentage of the structures of the shell can be determined, it has become possible to distinguish sharply between two kinds of selective pressures, those arising from the density-independent part of the environment and those arising from the density-dependent part of the environment. In the concluding section the significance of these two types of selective pressures is discussed.

POLYPHYLETIC EVOLUTION OF EGG PLASTRONS

In order to establish the independent evolution of similar organs or structures in two or more groups it is only necessary to show that such organs or structures are absent in the common ancestor of the groups concerned, that is, the organs or structures in question have no phyletic continuity. To demonstrate that two organs or structures are of independent origin is also to demonstrate that they are not homologous, however close and detailed may be their resemblance. Conversely, lack of resemblance between organs or structures of different groups is *of itself* no argument against homology because phyletic continuity may exist between organs or structures that have had a long history in quite different environments and may even have come to subserve quite dissimilar functions.

A natural classification of the different kinds of egg plastrons, that is, one that would reflect the phylogenetic relationships of the groups, is precluded on principle because of the polyphyletic origin of the plastrons. Egg plastrons have been placed in three groups (Hinton, 1961), but so many intermediates have been found that for the present it is probably best merely to speak of those with and without plastron-bearing horns. The vast majority of eggs that utilize plastron respiration lack plastron-bearing horns. In those lacking horns the plastron network may extend over the entire shell, as in most Muscinae, or it may be confined to a part of the shell. In the Diptera-Cyclorrhapha the plastron is often restricted to the area between the hatching lines, as in the eggs of many of the more specialized flies. The plastron is sometimes present only on one side of the eggshell, as in bugs of the family Saldidae (Cobben, 1968). Sometimes the plastron may consist of a few or many discrete islands or craters scattered over the surface of the shell, as in *Musca vetustissima* and *M. sorbens* (Plate III, Figs 1, 2).

I have recently (Hinton, 1969) drawn attention to the fact that according to the fine structure of the surface of the plastron two rather different types may

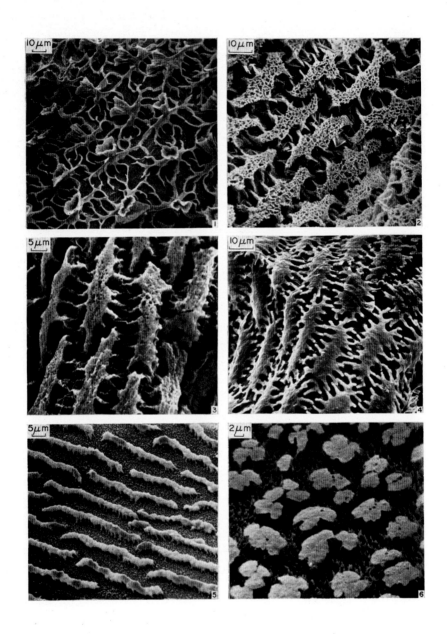

be recognized: (1) plastrons that clearly consist of an enlargement of the aeropyles (Plate IV, Fig. 3), or a great increase in their number, or both (Plate V, Figs 1, 2); and (2) plastrons such as those of the Muscidae that consist of a network (Plate II, Figs 1, 2). The origin of some surface networks by a sufficient approximation of many aeropyles is not difficult to visualize and no doubt often occurs. But it seems improbable that the plastron network of *Anopheles* eggs or those of *Chrysopa* (Plate II, Fig. 6) originated in this way. Aeropyles sufficiently numerous to form a plastron may be more or less evenly scattered over the surface, or they may be confined to certain restricted areas. Sometimes the aeropyles appear to be formed in the centres of hexagons that mark the boundaries of individual follicular cells, as in many long-horned grass-hoppers (Tettigoniidae). In many insects, on the other hand, the aeropyles tend to be confined to the boundaries of the hexagons, as they are in the eggs of many different kinds of moths (Plate IV, Figs 1, 2; Plate V, Figs 5, 6).

Egg plastrons have been evolved independently an enormous number of times, particularly within the orders Lepidoptera and Diptera. The great majority of instances of the independent origin of chorionic plastrons are to be found among eggs without respiratory horns, but as yet no attempt has been made to list these. Eggs with plastron-bearing horns are relatively rare (Figs 2 a–g), but, as I have recently shown (Hinton, 1969), there are no less than 19 groups in which such horns have been independently evolved, as follows:

HEMIPTERA

(1) Tettigometriidae (*Tettigometra*)
(2) Acanalonidae (*Acanalonia*)
(3) Tropiduchidae (some)
(4) Nepidae (Fig 2 e, g)
(5) Miridae (*Termatophylidea*)
(6) Miridae (some Cylapinae)

Plate I

Figs 1–4. Plastron network of side of shell of various species of Syrphidae.
Fig. 1. *Syrphidius ribsii* (L.).
Fig. 2. *Episyrphus balteatus* (Deg.).
Fig. 3. *Rhingia campestris* Meig.
Fig. 4. *Platycheirus peltatus* (Meig.).
Fig. 5. Side of shell of *Platypalpus pallidiventris* (Meig.), Empididae.
Fig. 6. Plastron of median area between hatching lines of *Polietes lardarius* (F.), Muscidae.

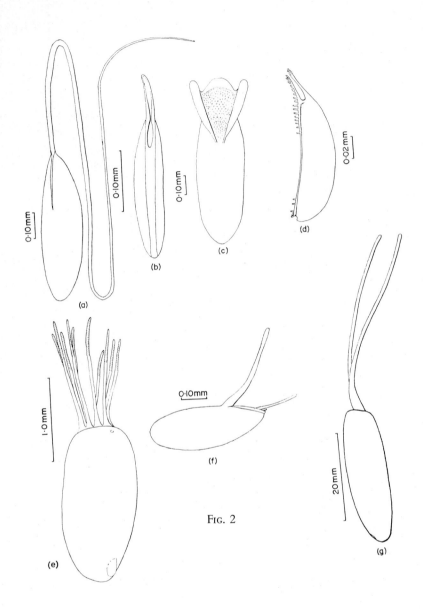

Fig. 2 (a, b, c, d, e, f, g). Eggs with plastron-bearing respiratory horns. (a) *Sepsis violacea* Meig., Sepsidae. (b) *Musca (Eumusca) autumnalis* Deg., Muscidae. (c) *Scopeuma stercorarium* (L.), Cordiluridae. (d) *Hebecnema umbratica* (Meig.), Muscidae. (e) *Nepa rubra* L., Nepidae. (f) *Drosophila gibberosa* (Patt. & Main.), Drosophilidae. (g) *Ranatra fusca* Beauv., Nepidae.

HYMENOPTERA
(7) Encyrtidae (some)

DIPTERA
(8) Dryomyzidae (*Dryomyza*)
(9) Sepsidae (many, Fig. 2a)
(10) Coelopidae (*Orygma*)
(11) Sphaeroceridae (*Coprophila*)
(12) Sphaeroceridae (some *Leptocera*)
(13) Drosophilidae (some *Drosophila* s.lat., Figs 1a and 2f)
(14) Cordiluridae (some, Fig. 2c)
(15) Muscidae (some *Eumusca*, Fig. 2b)
(16) Muscidae (*Myospila*)
(17) Muscidae (*Mydea*)
(18) Muscidae (*Hebecnema*, Fig. 2d)
(19) Muscidae (some *Limnophora*)

Plastron-bearing horns appear to be a primitive feature in only one family of insects, bugs of the family Nepidae (Hinton, 1961). All other families that contain species with plastron-bearing horns also contain numerous species without such structures. It may be noted that there are in fact more than 19 instances of independent origin in the groups cited above. For instance, I have cited the Drosophilidae as a single instance of the independent origin of such horns, but within the genus *Drosophila* horns have been evolved on several occasions unless we are to suppose that the common ancestor of recent species had a large number of horns and that differences in number between the species are due to reduction rather than to the acquisition of additional pairs of horns. Horns are certainly not a primitive feature in the Drosophilidae, and they are absent even in a few species of *Drosophila* (s.lat.) e.g. *D.* (*Phleridosa*) *flavicola* Sturt. and *D.* (*Hirtodrosophila*) *sexvittata* Okada. In the subgenus *Hirtodrosophila* there are species with and those without horns.

SOME OBSERVATIONS ON THE PRINCIPLES OF CLASSIFICATION

Statements about independent origins are statements about relationships, and their validity is necessarily dependent upon the degree to which nomenclature reflects phylogeny. In speaking of the independent evolution of structures, it has to be borne in mind that at any taxonomic level the preferred nomenclature is that which most nearly reflects the relative times of divergence of the different groups. Polyphyletic taxa are inevitably created when relative times of divergence are reversed. For instance, as I have shown elsewhere (Hinton, 1958), to

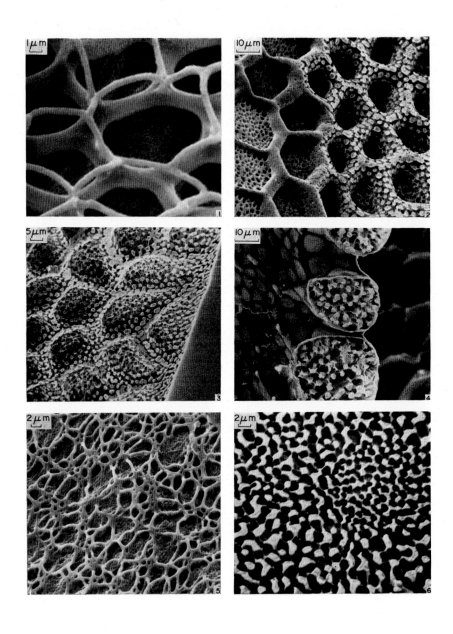

place the Permian Orthophlebiidae (or related families) in the order Mecoptera with recent Mecoptera is to create a polyphyletic order in terms of other panorpoid orders. To place the Orthophlebiidae in the Mecoptera is to say that they are more nearly related to the Mecoptera than to other panorpoid orders. In other words it is to say that they are more nearly related to the Mecoptera than are the Siphonaptera and Diptera to the Mecoptera, which is absurd. It must always be remembered that *one taxon can only be polyphyletic in terms of another, never in terms of itself*. In short, all living organisms could be placed in one family and this would not be polyphyletic although it would not be a useful classification because it would give no indication of phyletic relationships.

Mistakes in systems of classification arise essentially in three different ways: (1) because not enough facts are known about the taxa under consideration; (2) because the available evidence is misinterpreted, or; (3) because of a failure to understand the relation between nomenclature and the facts of evolution so that polyphyletic groups are created without intention. The first two kinds of mistakes do not require further explanation. Mistakes arising because of the failure to understand the relation between nomenclature and the facts of evolution can easily be avoided and have been much discussed since the 1950's by Hennig (1950) and others. Even so, there is still a widespread belief that phylogenetic systems should take into account the relative amounts of divergence from the primitive form, that the amount of divergence from the primitive form should be reflected in the nomenclature used. In other words, it is thought that differences in the rates of evolution should affect the position of the group in the nomenclatorial system. It is necessary to state most emphatically that in phylogenetic classifications we are in no way concerned with the amount of divergence. Concern with the degree or amount of divergence from the primitive plays so large a part in many supposedly phylogenetic classifications that it may be considered in a little more detail. In the first place, many of the ways chosen to portray phylogenetic systems, such as trees without a time scale,

---

Plate II

Fig. 1. Side of shell showing the coarse and fine plastron networks of *Fannia armata* (Meig.), Muscidae; the fine network is more resistant to wetting by high hydrostatic pressures than is the coarse network.

Fig. 2. Plastron of mesal side of wing of the eggshell of *Fannia coracina* (Loew).

Fig. 3. Plastron of side of wing of shell of *Fannia atripes* Stein.

Fig. 4. Same of *Fannia nidica* Collin.

Fig. 5. Plastron of ventral side of wing of shell of *Fannia canicularis* (L.).

Fig. 6. Plastron of side of shell of *Chrysopa* sp., Neuroptera.

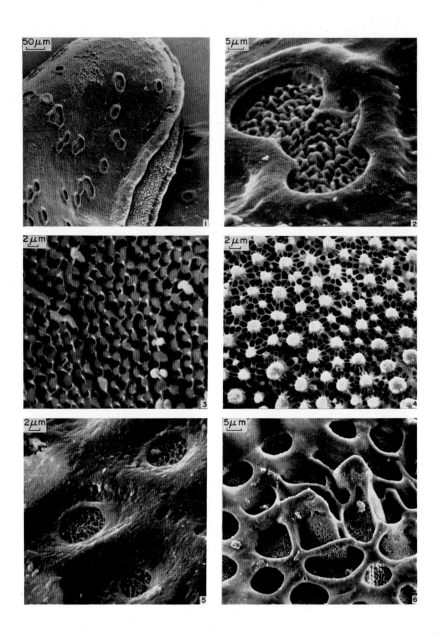

almost inevitably mean that allowance is being made for the amount of divergence from the common ancestor. However, it should be clear that if the amount of divergence of two or more groups from their common ancestor is allowed to influence the structure of a phylogenetic tree, it can only do so by altering time relations, that is, by reversing relative times of origin and therefore making polyphyletic groups.

The amount or degree of divergence from the primitive is an indication of the rate of evolution. Because the amount of divergence is an indication of the rate of evolution it has no necessary bearing upon the time of origin. For instance, the rates of evolution may be entirely different in very closely related groups. To consider but a single example, some insects enter ants' nests and in due course become what we call true guests or symphiles of the ants. There is a very great selective pressure on symphiles for a circumscribed range of structure. The very heavy selective pressure exerted upon the morphological features of a species that has become an ant guest means that its rate of evolution towards a different form and structure is vastly accelerated as compared with its relatives that remain outside ant nests. A particularly striking example of this is to be found in some ground beetles (Carabidae) of the tribe Ozaenini. In the early or middle tertiary one or more groups of Ozaenini became ant guests. Their rate of morphological change was so greatly accelerated that the Ozaenini found today in ant nests so little resemble their relatives outside that until recent years they were placed without question in a family of their own, the Paussidae. Thus in this instance the fact that the rate of evolution was allowed to affect classification resulted in making the Carabidae polyphyletic in terms of the family Paussidae. In short, polyphyletic categories are inevitably created when relative times of origin are reversed. In a phylogenetic classification no account can be taken of different rates of evolution because the rate of evolution is not related to the times of origin but to outward circumstance that has nothing to do with times of origin.

One of the most general laws of evolution is what I have called the law of unequal development (Hinton, 1958), borrowing a name from the 19th century

---

Plate III

Fig. 1. Anterior end of shell of *Musca sorbens* Weid., Muscidae.
Fig. 2. Plastron crater of same species.
Fig. 3. Plastron network of side of shell of *Orthellia caesarion* (Meig.), Muscidae.
Fig. 4. Plastron network of side of shell of *Culex pipiens fatigans* Wied., Culicidae.
Fig. 5. Plastron craters of side of shell of *Idiostatus aequalis* (Scudder), Tettigoniidae.
Fig. 6. Plastron craters around a micropyle of *Idionotus siskiyou* Hebard, Tettigoniidae.

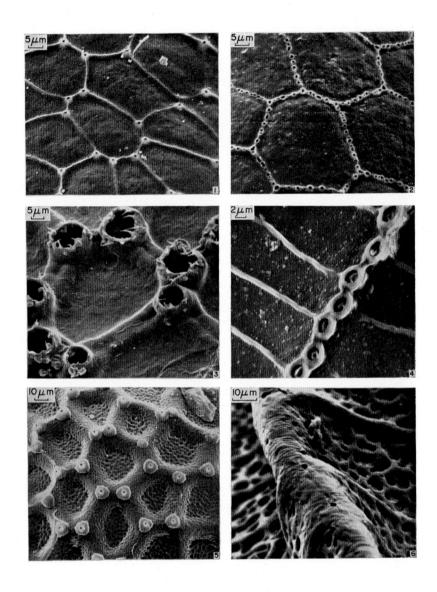

economists who were concerned with the evolution of different social systems. This law states that not only do different groups evolve at different rates but within any taxon the organ systems themselves evolve at different rates. This necessarily follows because the selective pressure of the physical and biological environment can never be expected to be exactly equal on every organ system for long periods of time. This is why so many animals and plants are such mixtures of specialized and primitive features. The idea of unequal development has been referred to a number of times in the work of D. M. S. Watson, and De Beer (1954) has referred to what I call the law of unequal development as "Watson's Rule".

Attempts are often made to assign values to different characters and to estimate the degree of specialization or the degree of primitiveness by summating these values. For instance, a primitive leg with the trochanter unfused to the femora is given, say, 5 points and one where the femora and trochanter are fused together is given, say, 0 points, and so on. When this has been done for a large number of structures or characters of one kind or another the total scores are then supposed to provide an answer as to which group is more specialized and which is more primitive. However, looking at the matter a little more closely, it will be seen that attempts of this kind are based upon a failure to understand that different morphological characters have no universal equivalent. They cannot be weighed one against the other any more than the use value of a shirt can be compared with the use value of a pair of shoes. The value of commodities with different use values can, however, be compared in terms of their value or price in money. That is, money functions as a universal equivalent for comparisons of this kind between different commodities but there is no universal

---

Plate IV

Fig. 1. Aeropyles of side of shell of *Arctia caja* (L.), Arctiidae.

Fig. 2. Aeropyles of side of shell of *Phragmatobia fuliginosa* (L.), Arctiidae; the total water-air interface of the aeropyles when the egg is flooded is about 1 to 2% of the surface, which in an egg of this size is sufficient to make plastron respiration of some significance.

Fig. 3. Aeropyles of side of shell of *Antheraea pernyi* Guér., Saturniidae; the aeropyles are greatly enlarged and the total area of the openings is about 12% of the surface of the egg, and there is therefore hardly any doubt that in this egg plastron respiration is significant when it is covered by water.

Fig. 4. Aeropyles of side of shell of *Ochropleura plecta* (L.), Agrotidae.

Fig. 5. Aeropyles of side of shell of *Semiothisa signaria dispuncta* Walk., Geometridae.

Fig. 6. Aeropyles on one of the longitudinal ridges of the eggshell of *Argynnis selene* (Schiff.), Nymphalidae.

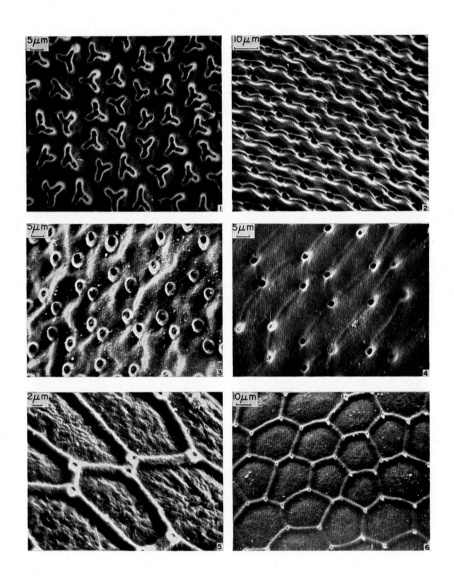

equivalent that can be used for comparing the value of morphological characters.

During the last decade there have been many papers published on what may be described as numerical taxonomy. In nearly all of these papers is the assumption, always implicit but occasionally explicit, that there is some kind of universal equivalence between characters. However, it seems clear enough that many pairs of taxa could converge in evolution to such an extent that 99% of all their characters were the same but the groups were nevertheless distantly related, that is they came from different stocks. Conversely, it often happens that two groups that share a common ancestor more recently than any other groups diverge so widely during the course of evolution that hardly any of their characters are the same. The numerical taxonomist has not yet invented a method of taking these circumstances into account. To put it another way: (1) the similarity might be due to convergence, i.e. the structures were not identical in all ancestors of the two groups *and* in the common ancestor of both, or (2) the present similarity of structures is due to the fact that they were possessed by the common ancestor of the two groups and in both have remained unchanged. If the first proposition is true, then the similarity, however great, should be given absolutely no weight in deciding relationships. If the second proposition is correct it is proof of a relationship. Most numerical taxonomists simply make lists of characters without distinguishing whether the first or second of these propositions applies to them, and it is difficult to see any value in the exercise apart from the fact that it is always useful to discover characters that may have been overlooked by others.

STRUCTURE IN RELATION TO DENSITY-DEPENDENT AND DENSITY-INDEPENDENT SELECTIVE PRESSURES

I am not aware that anyone has previously drawn attention to the bearing of structure upon the question of whether the total number of individuals of a

---

Plate V

Figs 1–6. Aeropyles of the side of the eggshell of various species of moths of the family Notodontidae.

Fig. 1. *Pheosia gnoma* (F.).
Fig. 2. *Pheosia tremula* (Clerk).
Fig. 3. *Ichthyura albosigma* Fitch.
Fig. 4. *Notodonta dromedarius* (L.).
Fig. 5. *Nadata gibbosa* (A. & S.).
Fig. 6. *Heterocampa manteo* Doubl. In the species of *Pheosia* the aeropyles are sufficiently numerous and large to constitute an effective plastron, whereas in *Nadata* and *Heterocampa* the aeropyles are too few and too small to constitute an effective plastron.

species existing at any one time are, by and large, determined by density-independent factors or by density-dependent ones. When the functional significance of the structures of an organism is understood, it becomes possible to decide whether most of the structures have been evolved as the result of selective pressures of a density-independent kind or as the result of selective pressures that can properly be called density-dependent. It is first necessary to digress a little and define terms in order fully to appreciate how the structure of an organism can assist in choosing between two quite different views about how numbers of individuals are regulated in nature.

By density-independent factors is meant those that affect the average individual of the species irrespective of population density. Climate is usually regarded as the most important density-independent factor. The quality and quantity of the available food are other density-independent factors.* Density-dependent factors, on the other hand, are those the effect of which on the average individual of a species depends upon the population density of that species. Intraspecific competition and predators, parasites, and pathogens are density-dependent factors. Intraspecific competition is the only density-dependent factor that is affected solely by numbers. Predators and parasites cannot be expected to show an unfailingly exact response to changes in the population density of their hosts because they have environmental relations that are quite independent of their hosts. The amplitude of the population density fluctuations of any species is regulated by a complex of factors which we speak of as the natural control of the species.

Broadly speaking, there are two sharply opposed schools of thought about natural control. Some writers claim that the total number of individuals of a species that are present at any one time depends only or chiefly upon density-dependent processes and they believe that density-independent factors do not limit density. Other writers assert that densities are determined by density-independent factors and sometimes go so far as to claim that it is not necessary to evoke density-dependent factors to explain either the maximal or minimal numbers of individuals of a species in nature. Long ago Milne (1957) pointed out that the density of each species fluctuates only within limits imposed by the total effect of the density-independent environment. My own view (Hinton,

---

* Competition for food is a density-dependent factor. Sometimes climate, which is ordinarily density-independent, is said to have a density-dependent action. For instance, when there are a more or less limited number of protected niches in the environment, individuals in excess of the number that can occupy the niches may be destroyed by unfavourable climate. But it is not the climate but the competition for the niches that is density-dependent.

1957) is that the fluctuations of pressure upon a species that result from variations in the total effect of the density-independent environment, especially climate, determine the fluctuations of its maximum potential density. But these potential densities are never attained because of (a) the action of imperfectly density-dependent factors such as parasites and predators, or, more rarely, because of (b) both the effect of imperfectly density-dependent factors and intraspecific competition. In short, the fluctuations of density permitted by the density-independent environment are nearly always dampened by imperfectly density-dependent factors, and only exceptionally is the combined effect of both so benign as to permit any significant degree of intraspecific competition.

Those who take the view that it is density-dependent factors that are chiefly responsible for natural control have a great deal of experimental evidence in support of their views, and it is therefore necessary briefly to examine the nature of the experimental evidence on which they rely. The total number of individuals of a species in nature is distributed as a greater or lesser number of more or less discrete communities. Each year, according to climate and other density-independent factors, some of these communities are wiped out and another year, when conditions are better, the sites where they previously existed are once more populated from not too distant communities that managed to survive.

In practically all experimental work a very small population or community of a species is kept in conditions where the density-independent factors are adjusted to levels at which the community multiplies, and density-dependent factors therefore become the important ones in stopping further increase. It is for this reason that so much attention is focused upon density-dependent factors, and the conclusions from such experiments are applied to the whole population of the species in nature. But it is quite clear that nobody setting up an experiment in a laboratory would adjust density-independent factors so that the population failed to multiply: to do so would appear absurd. The experiments are therefore done under conditions where no other conclusion is possible but that it is density-dependent factors that regulate numbers. In other words, the answer is predetermined by the nature of the experiment.

More rarely, a small community in nature is studied. Such communities have been studied over a short period of years, and even when something like 95% of the mortality is due to density-independent factors the writer nevertheless insists that these do not regulate density. However, work done with populations in nature is usually done only with extremely small natural communities that survive over a period of years simply because the density-independent factors have, during the period of observation, never been severe enough to wipe out the community. For instance, work has been done on the density fluctuations of a community on a few trees over a period of years when the climate was benign enough for the community to survive during the period of observation. But the species in question may occupy thousands or even millions of square miles. Conclusions based on a study of such small segments of the population are then treated as if they were relevant to the total population of the species. However, work on a small natural community is subject to precisely the same defects as the laboratory experiments. It is simply not possible, for the reasons already mentioned, to believe that the results from a study of a small community, surviving only where density-independent factors are benign, is necessarily relevant to what happens in nature where the total number of individuals of a species is distributed among a very large number of more or less independent communities, some of which survive and some of which are exterminated each year.

The relevance of all this to structure is simply that if the chief selective pressures on a species were density-dependent then we should expect to find this reflected in their structure or habits. If, on the other hand, the chief selective pressures are those resulting from density-independent processes we should expect to find that the chief variations in their structure were concerned with their relation with the density-independent part of their environment. To distinguish the effect of these two kinds of factors—density-dependent and density-independent—is particularly easy in the egg stage. This follows because in the egg stage the insect has to a large extent disrupted relations with the external physical and biological environment. Speaking loosely, the only demands that the insect egg makes upon its external environment are that it shall have an adequate supply of oxygen, not lose too much water, and that it shall be left alone. The insect in the egg stage has no social nor sexual behaviour, and it is thus structurally more simple than other stages where many of the structures are associated with social and sexual behaviour. Furthermore, the insect egg does not feed and is unable to move, which further makes for the simplicity of the shell and the possibility of apprehending the significance of its structure. When eggshells are examined from this point of view, it immediately becomes evident that the chief variations in structure between one species and another result from relations with the density-independent part of the environment. That is, they are concerned with the way in which the egg obtains its oxygen supply and conserves its moisture. Only relatively rarely are the variations in structure related to density-dependent factors such as parasites and predators. In fact, structures of the eggshell that are concerned with defence against parasites and predators, or against attack by larvae or adults of their own species, are so infrequent that they immediately command attention.

The enormous variation of sculpture to be seen in Plates I–V is related to the density-independent environment and in no way that I can see to the density-dependent environment. Nearly all the variation of structure reflects differences in the respiratory systems of the eggshell. These variations can be enormous even between related groups. For instance, Plate I, Figs 1–4 show the variation of the plastron of eggs of flies of the family Syrphidae. Plate II, Figs 1–4 show the enormous variation that can exist in closely related members of a single genus, in this case Muscid flies of the genus *Fannia*. The kind of variation to be found in a single family of moths is shown in Plate V, Figs 1–6. The primitive microsculpture of the outer surface of the eggshell consists of a pattern of polygons determined by the shape of the follicular cells that secreted the shell. This pattern is secondarily modified in many of the illustrations shown (Plate I, Figs 1–6; Plate IV, Figs 4 and 6; Plate V, Figs 1 and 2), but it does not appear that any of the secondary modifications here are related in any way to density-dependent selective pressures. Sometimes the boundaries of the polygons are elevated and the openings of the aeropyles are lifted well above the surface. The selective advantage of this is clearly that the egg is able to utilize atmospheric oxygen even when covered by a thin layer of water. It should also be noted that the structure of the respiratory system below the surface of the shell differs enormously even between closely related species: the chorion may have different numbers of layers, and the form of the internal chorionic meshworks can be very different.

Objections may be made to my thesis on the grounds that I have chosen a stage of the insect which has relatively uncomplicated relations with its physical and biological environment. Stages of the insect such as the larval stage or the adult stage with much more complex relations with the environment will undoubtedly show many more structures which are related to the density-dependent environment. However, a preliminary survey does not indicate that the *percentage* of structures evolved in relation to density-dependent selective pressures is any greater although the absolute number is of course much greater. I hope to be able to discuss this in more detail elsewhere.

SUMMARY

The geometry of the respiratory structures of insect eggshells has been surveyed with the aid of the scanning electron microscope. It has been shown that many structures that are identical in fine detail are polyphyletic in origin, and some of the principles of classification are considered in relation to polyphyletic origins of structures. The functional significance of most of the chief structures of eggshells has been determined. In so doing, it has become possible to distinguish sharply between two kinds of selective pressures, namely, those resulting from the density-independent part of the environment and those arising from the density-dependent part of the environment. The distinction that can be made between the two types of selective pressures has made it possible to assess the relative importance of these two kinds of selective pressures in the natural regulation of the numbers of individuals of a species, at least in so far as the egg stage is concerned.

REFERENCES

COBBEN, R. H. (1968). "Evolutionary trends in Heteroptera (Insecta, Hemiptera). Part I. Eggs, architecture of the shell, gross embryology and eclosion." *Medd.* 151, *Lab. Ent. Agric. Univ.*, Wageningen, Holland.

DE BEER, G. R. (1954). Archaeopteryx and evolution. *Advmt Sci., Lond.* **11**, 160–170.

HENNIG, W. (1950). "Grundzüge einer Theorie der phylogenetischen Systematik." Deutscher zentralverlag, Berlin.

HINTON, H. E. (1957). Biological control of pests. Some considerations. *Sci. Progr., Lond.* **45**, 11–26.

HINTON, H. E. (1958). The phylogeny of the Panorpoid orders. *A. Rev. Ent.* **3**, 181–206.

HINTON, H. E. (1961). How some insects, especially the egg stages, avoid drowning when it rains. *Proc. S. Lond. nat. Hist. Soc.* **1960**, 138–154.

HINTON, H. E. (1968). Spiracular gills. *Adv. Insect Physiol.* **5**, 65–162.

HINTON, H. E. (1969). Respiratory systems of insect egg shells. *A. Rev. Ent.* **14**, 343–368.

MILNE, A. (1957). The natural control of insect populations. *Can. Ent.* **89**, 193–213.

3 | Scanning Electron Microscopy of the Calcareous Skeleton of Fossil and Living Brachiopoda

ALWYN WILLIAMS

Department of Geology, The Queen's University, Belfast, Northern Ireland

Abstract: Studies of living and fossil calcitic-shelled Brachiopoda show that exoskeletal growth has never involved more than a few secretory processes. Variation in the periostracum of living species suggests that the prototypic organic covers were an impersistent mucopolysaccharide layer underlain by a fibrillar, proteinous, triple-unit membrane. The mineral exoskeleton beneath the periostracum of most Articulata has always consisted of a primary layer of highly inclined crystallites and a secondary layer of interlocking and alternating calcitic fibres sheathed in protein sheets. Minor changes include the neotenous suppression of the secondary layer in the Thecideidina and the gerontomorphic introduction of a prismatic tertiary layer in the Koninckinacea. A major divergence gave rise to the exoskeleton of the Strophomenida by the paedomorphic substitution of laminar shell secretion for most of the normal primary and secondary shell. The structure of the lath-like laminae suggests that the strophomenide shell grew by alternating secretion of protein and carbonate. This is quite different from the simultaneous secretion of protein and calcite during the spiral growth of the plate-like laminae forming the secondary shell of the inarticulate Craniacea, Craniopsidae and Obolellida, all of which evolved independently of each other.

INTRODUCTION

Electron microscopy promises to revolutionize palaeontological research. The use of techniques, hitherto the prerogative of the cell biologist, is now beginning to enlarge our understanding of the ultrastructural organization and growth of fossils. This greater comprehension is, in turn, leading to the discovery of new ways of determining the course of evolution, the validity of current classifications, and the nature of past environments. In no phylum is such progress better illustrated than in the Brachiopoda with their unsurpassed record of occurrence as members of marine benthic faunas from the early Cambrian to the present day. The brachiopod shell is noteworthy for the compositional as well as the structural changes it has undergone since the emergence of the phylum. These differences in composition transgress the systematic boundaries of the two brachiopod Classes erected on anatomical and embryological distinctions. Most Inarticulata have always had a chitinophosphatic shell, but a few groups are characterized by a protein-calcitic shell and it is probable that one extinct stock, the Trimerellacea, had a protein-aragonitic exoskeleton (Jaanusson, 1966). In contrast, the shell of the Articulata has always been exclusively protein-calcitic, although subject to much greater textural variation than that of the Inarticulata. So far, most of the publishable research has been carried out on the calcareous-shelled brachiopods and forms the basis of this review. Current fabric studies of the chitino-phosphatic skeleton, however, may radically alter our ideas on relationships within and between the Inarticulata and Articulata.

The procedure for investigating fossil skeletons is governed by the Huttonian principle of uniformitarianism, i.e. the present is the key to the past. In living brachiopods, the skeletal fabric can be related to the external form and secretory activity of cells composing the outer epithelium of the mantle lining the shell. Recent calcareous-shelled species belong to three (or possibly four) Orders which first appeared in middle or early Palaeozoic times. Comparative studies of their ancestral stocks will, therefore, show whether any changes in skeletal structure (and, by inference, in cell biochemistry) occurred during the history of the Orders. Having established reliable standards based on these Orders, it then becomes possible to determine whether the shell fabric of any extinct stock from the same locality as ancestors of living species is natural or whether it has been significantly altered by such *post-mortem* changes as may accompany consolidation of the entombing sediments. Usually, alteration of the original shell fabric by diagenesis and lithification becomes greater with increasing geological age. Yet it is still possible to recover early Cambrian shells with fabrics sufficiently well preserved for ultramicroscopic comparison with those of living forms. The reasons for such exceptional preservation are not well understood. The periostracum, as long as it survives, and the outer primary layer, which is normally recrystallized even in fossils collected from comparatively young unaltered sediments, clearly afford some protection for the rest of the shell. The segregation of the secondary layer into fibres or laminae, each enclosed in an organic membrane, may also have helped to retard recrystallization and reduce deformation of that layer. It is even possible that all shells became recrystallized in time but that individual fibres or laminae, constrained by organic membranes or their residues, recrystallized independently of one another into structures not differing much from the original crystallographically or morphologically. Whatever the reason, standard successions of organic and mineral layers, each the product of a particular secretory régime, can be established for each Order.

The nature of organic constituents associated with shells of extinct species and their location within the shell have usually to be inferred, although organic residues, identifiable as amino-acids, survive within Silurian articulate shells (Jope, 1967), and the protegula of early Cambrian Acrotretida bear microscopic moulds of the periostracum (Biernat and Williams, 1970).

The phylogenetic relationships revealed by such data have already warranted a reappraisal of the affinities of the Palaeozoic Triplesiidina and the Mesozoic Cadomellacea and Koninckinacea, with correspondingly important revisions of the ordinal status and evolutionary history of these groups (Williams, 1968a). Other revisions of equal weight are known or forthcoming. Moreover the fabric of living shells includes ultramicroscopic interfaces marked by changes in either the composition or the rate of deposition of materials secreted by outer epithelium. These surfaces recur with periodicities indicative of diurnal deposition. Traces of them in fossil shells afford the prospect of being able to estimate absolute time and identify seasons by fabric analyses of fossil assemblages. Both estimates can reveal much about the environments that accommodated the faunas.

Once decalcified tissue has been examined under the transmission electron microscope and the relationship between mantle and shell established, research can be concentrated on the fabric of living and fossil exoskeletons. For this purpose the scanning rather than the transmission electron microscope is the more effective instrument. Material is more easily prepared for examination under a scanning beam and significantly larger surface areas can be scrutinized in one operation. The latter advantage and the wide range of magnification also ensures that the least altered patches of a specimen can be precisely and quickly located. The relatively low resolution of the scanning electron microscope is no handicap in this kind of work. The relief and crystallography of prepared surfaces are easily decipherable at resolutions of about 20 nm which is within the range of current scanning models. In effect, the versatility of the scanning electron microscope more than outweighs its disadvantages in the exploration of ultramicroscopic topographies as the results of the researches outlined below will show.

MATERIALS AND METHODS

Ultrastructural investigations of the brachiopod exoskeleton necessitate the use of many materials ranging from recrystallized fossils embedded in various rocks to unaltered shells of living and extinct species that are free of matrix.

Sections of soft parts are prepared by fixing living specimens in 3% gluteraldehyde made up in 3% sodium chloride. The brachiopods are then decalcified in 10% EDTA, washed in sucrose and treated for 1 hour with 2% osmic acid;

all solutions being buffered to pH 7·2 with phosphate buffer. Following dehydration, specimens are embedded in an Epoxy resin and the sections stained with aqueous uranyl acetate and aqueous lead citrate for examination under the EM6B transmission electron microscope.

Natural and fracture surfaces of the shells of living brachiopods are first cleaned for some hours in sodium hypochlorite to break down any associated tissue and then briefly vibrated in acetone in an ultrasonic tank to remove adherent organic debris. Surfaces of free fossil shells usually have first to be cleaned of adherent patches of clay or sand matrix. This can be done by loosening the matrix with a needle and then subjecting the specimen, immersed in detergent, to ultrasonic agitation.

The majority of fossil specimens examined are embedded in rock matrix. Indeed, more complete and less altered sections of the primary layer and the inner parts of the secondary layer are likely to be provided by shells enclosed in sediment than by free specimens which have almost invariably suffered some degree of superficial weathering. However, fossils in indurated matrix can rarely, if ever, be successfully prepared to show the true ultrastructure of natural and fracture surfaces, and the constituents of their skeletal successions are best seen in cut sections. Consequently, Recent shells have also to be sectioned in the same way to ensure that the structural elements of fossil successions have been correctly identified. For this purpose, both fossils in matrix and free shells are embedded in a resin like "Araldite" that can be polymerized to form transparent blocks. A block can then be cut with a diamond-edged blade along any preferred direction, and surfaces, containing the required sections, can be polished with tin oxide or polishing alumina and etched in 2% EDTA for between 15 and 30 min dependent on the texture of the shell and the compaction of the matrix.

All natural surfaces, fracture surfaces and differentially etched sections are coated with gold-palladium for study under the scanning electron microscope (Cambridge Scientific Instruments Ltd Stereoscan).

SHELL SUCCESSION OF ARTICULATA

1. Rhynchonellida and Terebratulida

Both the Rhynchonellida and the Terebratulida are represented by widely distributed species in modern seas and are characterized by fundamentally similar shell successions. However, because the impunctate shell of the Rhynchonellida is simpler in arrangement and because that Order first appeared in the Ordovician long before the earliest Terebratulida of Devonian age, the shell succession of the living rhynchonellide, *Notosaria nigricans* (Sowerby), is taken as standard (Williams, 1968b).

In *Notosaria*, as in all living brachiopods so far examined, proliferation of cells making up the mantle secreting a valve is maintained by a circumferential groove situated just within the mantle edge (Fig. 1). As each cell is released from the generative zone, it displaces its peripheral neighbour so that every cell rotates in turn around the outer lobe or tip of the mantle to become an integral part of the outer epithelium underlying the shell. During rotation, the secreting plasmalemma of each cell performs six depositional operations in strict sequence, which constitute the standard secretory régime.

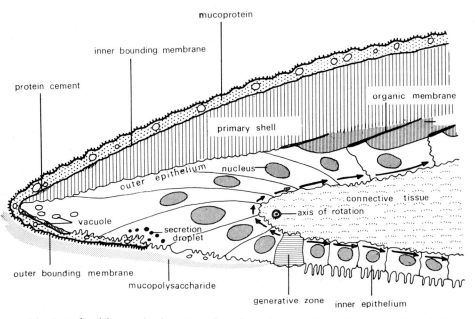

Fig. 1. Stylized longitudinal section of a valve of young *Notosaria nigricans* (Sowerby) showing the outer mantle lobe in relationship to the periostracum and the calcareous shell.

As a cell moves away from the generative zone it first secretes a mucopolysaccharide layer which rarely persists beyond the shell edge. Beneath the mucopolysaccharide, the cell then assembles a triple-unit membrane, mainly proteinous in composition and about 14 nm thick with an external array of protein rods about 20 nm long. The membrane forms an outer cover to the chief constituent of the periostracum, a mucoprotein layer up to 1 μm thick which, when fully developed, contains long fibrils and sporadically distributed vesicles and mucin droplets. This layer is then sealed by another triple-unit membrane also bearing distal extensions in the form of a dense mat of fibrils.

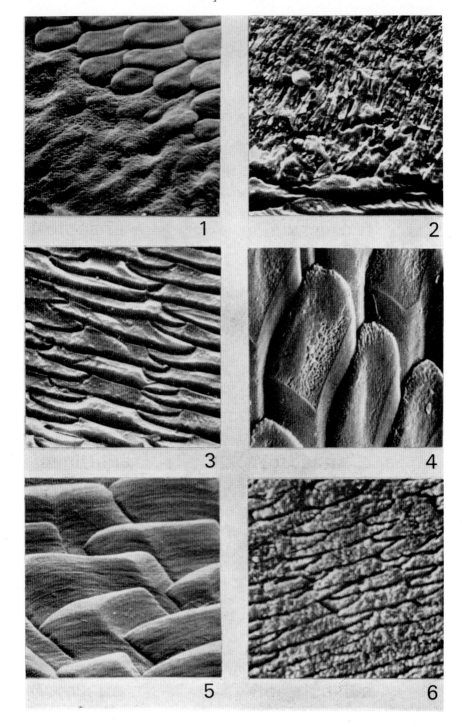

Plate I

Fig. 1. View of the internal surface of a valve of *Notosaria nigricans* (Sowerby) showing the junction between the primary layer of fine crystallites and the secondary layer of alternating, overlapping fibres; edge of valve beyond bottom left hand corner of micrograph.
× 1250.

Fig. 2. Etched section of a valve of *Notosaria nigricans* (Sowerby) showing the junction between the primary layer of fine, sporadically banded crystallites and blade-like fibres of the secondary layer; exterior of valve beyond top edge of micrograph.
× 2600.

Fig. 3. Etched section of a valve of *Hemithiris psittacea* (Gmelin) showing the characteristic transverse outline of fibres of the secondary layer; interior of valve beyond bottom edge of micrograph.
× 2400.

Fig. 4. View of the internal surface of a valve of *Notosaria nigricans* (Sowerby) showing the mosaic of the secondary layer made up of exposed terminal faces of fibres; tips of fibres point antero-medianly.
× 2400.

Fig. 5. View of the internal surface of an adductor scar of the brachial valve of *Notosaria nigricans* (Sowerby) showing modified terminal faces of fibres beneath muscle bases; truncated edges of fibres facing anteriorly.
× 2400.

Fig. 6. Etched section of a valve of *Rostricellula lapworthi* (Davidson), Upper Ordovician, Scotland, showing the characteristic transverse outline of fibres of the secondary layer; interior of valve beyond bottom edge of micrograph.
× 2600.

By the time a cell completes its portion of the periostracum, it occupies the tip of the outer mantle lobe. As it rotates around the tip the secreting plasmalemma of the cell which now faces outwards in the characteristic attitude of the outer epithelium, deposits calcite rhombs on an embedding protein cement forming the inner surface of the periostracum. With continuing carbonate deposition the calcite nuclei grow laterally and vertically and unite into a continuous layer. This is the primary calcareous layer of the brachiopod shell. (Plate I, Fig. 1). It is composed of amalgamated crystallites commonly finely banded at intervals of about 300 nm (Plate I. Fig. 2). The banding results from periodic variation in cell secretion although organic materials are rarely deposited and never in regular alternation with carbonate. However, when a cell, through addition of newer cells at the tip, comes to occupy a given distance behind the tip, it reverts to the exudation of a definite organic membrane as well as continuing carbonate secretion (Plate I, Fig. 1). The membrane, which is a triple-unit structure about 14 nm thick, is exuded by the microvillous anterior arc of the plasmalemma while the calcite is secreted from the posterior part. Both deposits constitute the secondary shell and, because the outer epithelial cells are regularly arranged in alternating rows, the membranes join up with one another to segregate the calcite into long thin rods known as fibres. The fibres tend to be blade-like in longitudinal section (Plate I, Fig. 2), but in transverse section they have a highly characteristic outline of inwardly convex curves representing the regular stacking and interlocking of fibres (Plate I, Fig. 3). The fibres are usually inclined at about 10 degrees to the primary shell but may twist spirally or tilt at all angles during subsequent growth. The terminal faces of fibres form the internal surface of the valve. Their peripheral boundaries are usually outwardly convex arcs overlapping one another in regular alternating rows in a distinctive pattern known as the mosaic. The growth of fibres is controlled at the exposed terminal face corresponding to the back part of a cell, and normally identifiable as a pitted area with boundaries subtending a rhombohedral angle (Plate I, Fig. 4). The terminal faces and, therefore, the cross sections of fibres can change in proportions according to variations in the relative rates of calcite and membrane deposition. But so far, gross modification involving the non-deposition of membranes and the amalgamation of adjacent fibres, has been found only in parts of the shell underlying muscle bases (Plate I, Fig. 5). Resorption can also obscure the secondary mosaic, but the process is restricted to certain features, like the cardinalia and lophophore supports, that are continually changing shape.

The standard rhynchonellide succession therefore comprises from the exterior inwardly: a mucopolysaccharide layer, a periostracum consisting of two triple-

3. SEM Studies of Skeletons of Brachiopoda

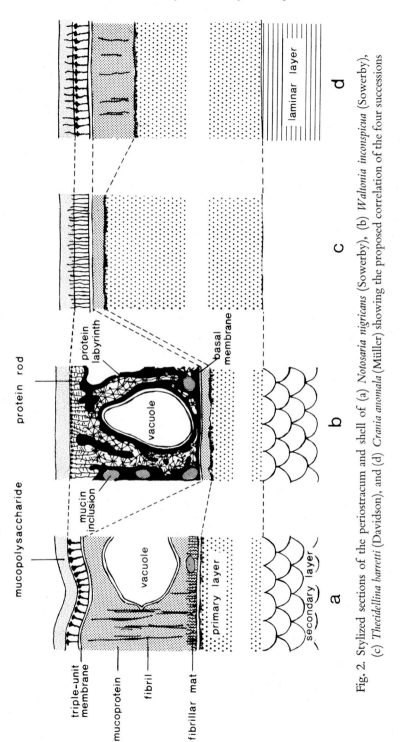

Fig. 2. Stylized sections of the periostracum and shell of (a) *Notosaria nigricans* (Sowerby), (b) *Waltonia inconspicua* (Sowerby), (c) *Thecidellina barretti* (Davidson), and (d) *Crania anomala* (Müller) showing the proposed correlation of the four successions

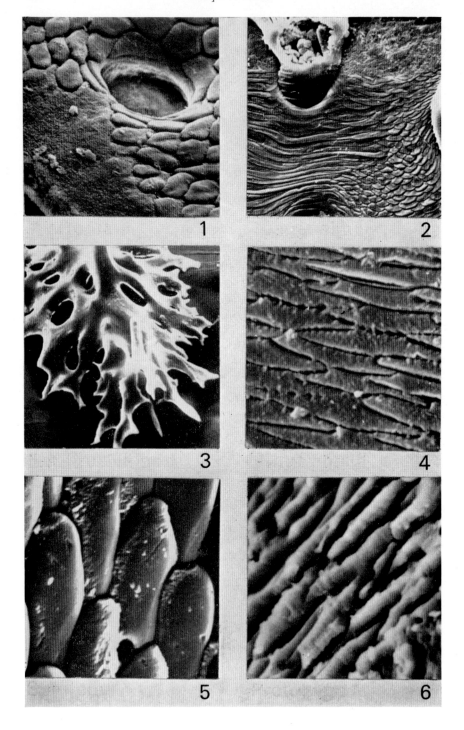

Plate II

Fig. 1. View of the internal surface of a valve of *Waltonia inconspicua* (Sowerby) showing the junction between the primary layer of fine crystallites and the secondary layer of alternating, overlapping fibres almost surrounding a puncta; edge of valve at bottom left hand corner of micrograph.
×1200.

Fig. 2. Etched section of a valve of *Magasella sanguinea* (Leach) showing punctae piercing the primary and secondary layers with resin infills of canals of a canopy simulating a brush in the top left hand edge which also marks the exterior of the valve.
×600.

Fig. 3. View of part of a spicule from the mantle of *Terebratulina retusa* (Linnaeus).
×250.

Fig. 4. Etched section of a valve of *Mutationella podolica* (Siemiradzki), Siluro-Devonian Czortkow Beds, Poland, showing the characteristic transverse outline of the fibres of the secondary layer; interior of valve beyond bottom edge of micrograph.
×2700.

Fig. 5. View of internal surface of a valve of *Spiriferina walcotti* (Sowerby), Lower Jurassic, England, showing a mosaic made up of exposed terminal faces of secondary fibres; tips of fibres point antero-medianly.
×1300.

Fig. 6. Etched section of a valve of *Protozyga rotunda* Cooper, Upper Ordovician, Virginia, showing the characteristic transverse outline of fibres of the secondary layer; interior of valve at bottom right hand corner of micrograph.
×2400.

unit membranes bounding a mucoprotein layer, a primary layer composed of cryptocrystalline or sporadically banded calcite, and a secondary layer consisting of calcite fibres sheated in protein membranes and stacked in alternating rows (Fig. 2a). The organic elements of the succession have not been visually identified in any fossil species. But the primary and secondary calcareous layers are recognizable even in the oldest known rhynchonellide, the Ordovician *Rostricellula* (Plate I, Fig. 6), and it is reasonable to assume that the entire succession of living species has always been typical of the Order. One reservation has to be borne in mind. As comparison with the periostracum of the Terebratulida and other Orders emphasizes, the rhynchonellide periostracum is exceptional in having two triple-unit membranes both with distal fibrillar extensions (Fig. 2). Rhythmic layering involving the repetition of the mucopolysaccharide layer and the outer triple-unit membrane can occur during deposition of the *Notosaria* periostracum. It is therefore possible that the modern structure is a genetically stabilized repetition of what was a single bounding membrane in ancestral stocks, and that the prototypic periostracum of articulate brachiopods was a fibrillar triple-unit membrane underlying an impersistent layer of mucopolysaccharide.

The calcareous succession of the Terebratulida is similar to that of the Rhynchonellida (Brunton, 1969). Both the primary layer and the secondary layer with its characteristically stacked fibres have always been well developed (Plate II, Figs 1, 2). The layers, however, differ in being penetrated by canals punctae, presumably of the same design, are invariably present in all fossil and tae (Plate II, Fig. 1) are separated from the periostracum by sieve-like canopies of primary shell about 1 μm thick, which contain tubular extensions (brush) of the membrane lining the punctae (Owen and Williams, 1969). Although punctae, presumably of the same design, are invariably present in all fossil and living Terebratulida, they are not unique to that Order but occur in three others including the Rhynchonellida (the late Palaeozoic *Rhynchopora*).

The full skeletal succession of the Terebratulida differs from that of the Rhynchonellida in two respects other than punctation. The periostracum consists of an array of level-topped protein rods connected by fibrils. The rods arise from a labyrinth of protein partitions enclosing vesicles and mucin droplets, which is really an extension of the outermost layer of a triple-unit basal membrane. Both vesicles and droplets provide the periostracum with a conspicuous external topography. The most reasonable correlation between the rhynchonellide and terebratulide periostraca equates the outer triple-unit membrane of the former with the basal triple-unit of the latter (Fig. 2b).

The other important difference is that a true endoskeleton is developed in many Terebratulida. It consists of a loose or continuous meshwork of spicules (Plate II, Fig. 3) secreted by scleroblasts within the connective tissue of the mantle and lophophore (Williams, 1968b). Spicules are known to occur in fossil Terebratulida as well as fossil and Recent Thecideidina and might have been characteristic of other groups now extinct.

An examination of fossil Terebratulida shows that at least the mineral constituents of the shell have remained unchanged since the time when *Mutationella*, the earliest known representative of the Order, flourished (Plate II, Fig. 4), and the full exoskeletal succession found in living species is regarded as typical.

2. Other Articulate Brachiopods

Except for the Thecideidina which first appeared in the early Mesozoic and survive today, all remaining articulate brachiopod groups are extinct and are known only by the calcareous part of their exoskeleton. In the living thecideidine, *Thecidellina*, the exoskeletal succession is reduced to: an external impersistent mucopolysaccharide layer, a fibrillar triple-unit membrane with a greatly thickened inner unit, and an internal undifferentiated calcareous layer (Fig. 2c). The calcitic layer, which is punctate, consists of crystallites disposed more or less normal to growth surfaces. The layer also frequently bears a fine banding with a periodicity of about 400 nm (compare Plate III, Fig. 2) which is unlike the lamination of the Strophomenida but identical with that found in the primary shell of living Rhynchonellida and Terebratulida. This likeness and the absence of secondary fibres suggest that the entire calcitic layer of living Thecideidina is primary shell. The development is noteworthy because in some Mesozoic stocks like *Moorellina*, a vestigial secondary fibrous layer still persists (Baker, 1970).

The structure of the thecideidine periostracum may also have an evolutionary significance. The entire periostracum beneath the external mucopolysaccharide layer correlates with only the outer and basal triple-unit membranes of the Rhynchonellida and Terebratulida respectively (Fig. 2). It is the simplest organic cover known in living brachiopod species and is provisionally assumed to have been characteristic of all extinct articulate Orders.

The youngest indisputable Spiriferida, the Jurassic *Spiriferina*, is indistinguishable in its punctate mineral shell structure (Plate II, Fig. 5) from the typical terebratulide (D. MacKinnon, personal communication). Both primary and secondary layers are orthodoxly developed in the majority of Spiriferida, punctate and impunctate (Armstrong, 1969) including the earliest known stock, the mid-Ordovician *Protozyga* (Plate II, Fig. 6). A normal calcareous succession is also characteristic of the impunctate Pentamerida and has been confirmed even in Cambrian porambonitaceans like *Bobinella* (Plate III, Fig. 4), although the

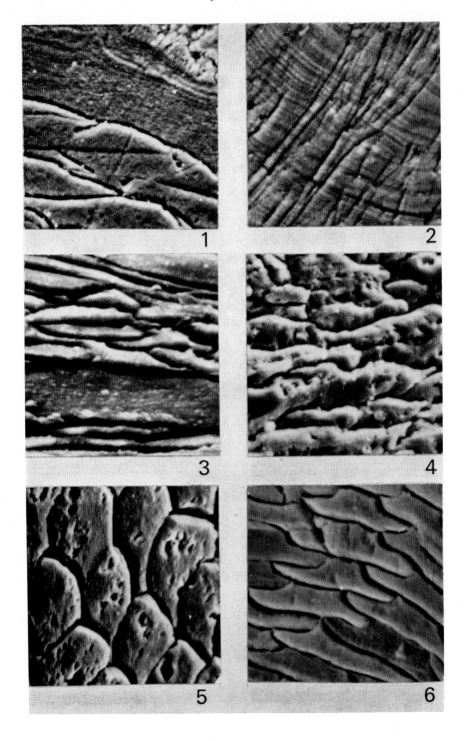

Plate III

Fig. 1. Etched section of a valve of *Koninckina leonhardi* (Wissmann), Trias, Austria, showing the junction between secondary layer and the fibrous, banded, prismatic tertiary layer; interior of valve at top right hand corner of micrograph.
×1200.

Fig. 2. Etched section of a valve of *Lacazella mediterranea* (Risso) showing depositional banding; interior of valve beyond bottom left hand corner of micrograph.
×2600.

Fig. 3. Etched section of the secondary layer of a valve of *Pentamerella* cf. *lingua* Imbrie, Middle Devonian, Michigan, showing stacks of fibres with normal transverse and oblique outlines alternating with prismatic calcite bands; exterior of valve beyond top edge of micrograph.
×2400.

Fig. 4. Etched section of the secondary layer of a valve of *Bobinella kulumbensis* Andreeva, Upper Cambrian, Vostochnaya, Siberia, showing the characteristic transverse outline of fibres; exterior of valve beyond top left hand corner of micrograph.
×1200.

Fig. 5. View of the internal surface of a valve of *Rhipidomella* sp., Pennsylvanian, Texas, showing the mosaic of exposed terminal faces of secondary fibres; tips of fibres point antero-medianly.
×1300.

Fig. 6. Etched section of the secondary layer of a valve of *Rhipidomella* sp., Pennsylvanian, Texas, showing the characteristic transverse outline of fibres; exterior of valve beyond top edge of micrograph.
×2400.

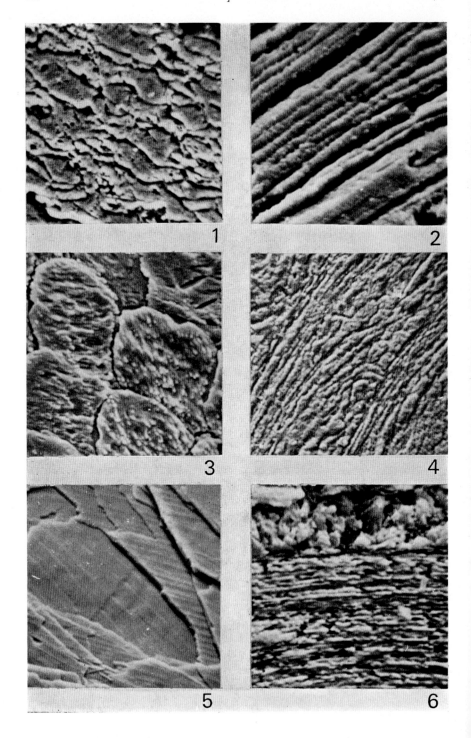

Plate IV

Fig. 1. Etched section of the secondary layer of a valve of *Nisusia ferganensis* Andreeva, Lower Cambrian, Ferganskaya Dolina, U.S.S.R., showing the characteristic transverse outline of fibres affected by recrystallization; exterior of valve beyond top right hand corner.
× 1200.

Fig. 2. Etched section of a valve of *Billingsella lindstromi* (Linnarsson), Middle Cambrian, Sweden, showing the laminar nature of the partly recrystallized secondary layer; exterior of valve beyond top left hand corner.
× 2400.

Fig. 3. View of the internal surface of a valve of *Sowerbyella variabilis* Cooper, Upper Ordovician, Oklahoma, showing the mosaic of exposed terminal faces of secondary fibres; tips of fibres pointing anteriorly.
× 1200.

Fig. 4. Etched section of a valve of *Bimuria* cf. *buttsi* Cooper, Upper Ordovician, Scotland, showing laminae in longitudinal section passing proximally (top right hand corner) into recrystallized nodes of normal primary shell; anterior of valve beyond top left hand corner.
× 650.

Fig. 5. View of exfoliated internal surface of *Schuchertella haraganensis* Amsden, Lower Devonian, Oklahoma, showing calcitic sheets formed of amalgamated laminae.
× 1200.

Fig. 6. Etched transverse section of a productidine spine, Lower Carboniferous, N. England, showing a thin primary layer (top part of micrograph) and underlying laminar layer.
× 2600.

fibrous secondary layer in both Orders was repeatedly modified by the development of lenses or layers of prismatic calcite (Plate III, Fig. 3) which was continuous enough to form a tertiary layer in at least one group of Spiriferida, the Koninckinacea (Plate III, Fig. 1).

The Orthida which became extinct in the Permian and include the earliest known articulate brachiopods were less uniform in shell structure. The Enteletacea (Plate III, Figs 5, 6) are punctate and their shell structure is identical with that of the Terebratulida although the presence of distal sieve-like canopies covering the punctae have yet to be seen. The Orthacea are impunctate and their mineral succession is indistinguishable from that of the Rhynchonellida so that a secondary mosaic can be identified even in Upper Cambrian *Orusia*. In contrast, the sketal successions of species currently assigned to the Billingsellacea appear to belong to two basically different types. In early Cambrian *Nisusia* (Plate IV, Fig. 1) and *Kotujella*, both a normal fibrous secondary layer and a primary layer, overprinted by traces of punctation in the latter stock, can be recognized. The secondary layer of *Billingsella lindstromi* (Linnarsson), however, is composed of flat-lying plates, about 1 μm thick (Plate IV, Fig. 2), which are like the laminae found in the Strophomenida and calcareous-shelled Inarticulata. The relationship between this anomalous species and the other early Palaeozoic Articulata is still being investigated.

The shell structure of the Strophomenida is fundamentally different from that of other Articulata. The shell of nearly all strophomenide species is penetrated by rods of coarsely crystalline calcite (pseudopunctate condition) which stand above the internal surface as small tubercles and which must have been secreted by regularly proliferated patches of specialized outer epithelium. In the oldest strophomenide group, the Plectambonitacea, the secondary layer consists of normally developed fibres (Plate IV, Fig. 3). The primary layer is seldom preserved, but in *Bimuria*, where it is excessively developed, it consists mainly of lath-like calcitic laminae regularly stacked parallel to the shell exterior (Plate IV, Fig. 4). Each lamina is about 2 μm thick distally but passes proximally into a much thicker crystalline lens with traces of vertically stacked crystallites. The arrangement suggests that the plectambonitacean primary layer was deposited by a retractible mantle with the proximal lenses secreted slowly immediately after retraction of the mantle edge and the laminae secreted quickly following mantle extension. In all other Strophomenida (including the Triplesiidina) no secondary fibres were deposited and the shell succession consists solely of laminae (Plate IV, Fig. 5) with vertically disposed crystallites forming an external layer about 6 μm thick in some productaceans at least (Plate IV, Fig. 6). Both layers correlate with the primary shell of the Plectambonitacea although the external crystallite skin is, strictly speaking, the primary layer, and the laminar shell, the secondary layer. The laminae are normally ribbed by sets of faint parallel ridges disposed at acute angles to one another in successive laminae (Plate IV, Fig. 5) (Towe and Harper, 1966; Williams, 1968a; Armstrong, 1969). The spacing of ridges corresponds to dimensions of rectangular mosaics exceptionally found in the internal surfaces of the best preserved specimens. They could therefore represent either impressions of cell boundaries or traces of protein thongs because organic material in some form or other separated the laminae from one another as is shown by the recovery of amino acids from a productid shell (Jope, 1967).

The strophomenide fabric is the most important departure from the standard articulate succession of a cryptocrystalline primary layer and a fibrous secondary layer. It is so distinctive that its occurrence in the Triplesiidina prompts the removal of that suborder from the Orthida to the Strophomenida.

SHELL SUCCESSION OF CALCITIC INARTICULATA

The Acrotretida, one of the four orders unequivocally assigned to the Inarticulata, is represented in modern seas by species with either chitino-phosphatic or protein-calcitic shells. The calcareous species, like the common *Crania anomala* Müller, belong to the Craniacea which, except for two Ordovician genera, have always lived with the pedicle valve cemented to the substrate (Williams and Wright, 1970). The cementing medium is the mucopolysaccharide film exuded by outer epithelial cells newly released from the intramarginal generative zone of the mantle. In the brachial valve, the mucopolysaccharide layer simply forms the external cover to a fully developed shell succession. Here, the periostracum consists of a mucopolysaccharide-mucoprotein layer, permeated by fibrils and up to 5·5 μm thick, which is bounded externally by a triple-unit membrane about 15 nm thick and internally by a proteinous monolayer. The triple-unit membrane is like the outer one of living Rhynchonellida because it, too, bears a distal array of proteinous rods with fibrillar extensions. The inner sealing membrane acts as a seeding sheet for the calcitic primary layer. This layer is composed mainly of acicular crystallites usually about 150 nm thick and inclined at about 45 degrees to isochronous surfaces (Plate V, Fig. 1). The crystallites show growth banding and may amalgamate into impersistent lenticles of relatively massive calcite up to 5 μm thick. Simultaneous secretion of organic compounds by outer epithelium is sporadic and restricted to the deposition of impersistent protein strands.

As in the Articulata, secretion of the primary layer continues until the outer epithelial cells, through the addition of new cells at the expanding mantle margin, are at a given distance from the edge of the valve. When this position is

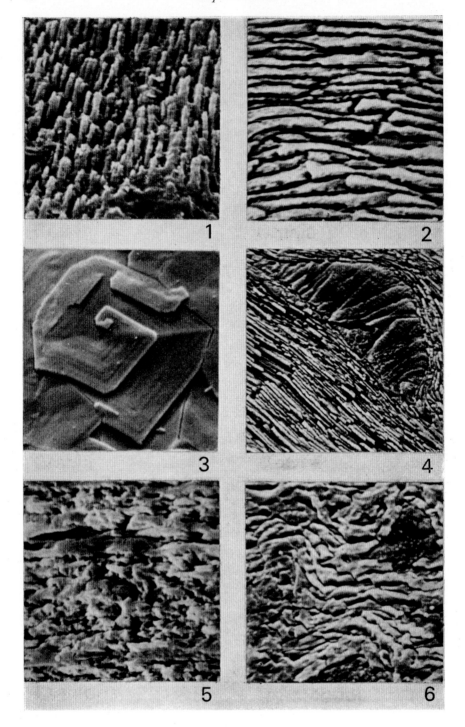

Plate V

Fig. 1. View of internal surface of a brachial valve of *Crania anomala* Müller showing acicular crystallites of the primary layer in various stages of amalgamation; tips of crystallites point towards valve margin.
× 4000.

Fig. 2. Etched section of the secondary layer of a brachial valve of *Crania anomala* Müller showing a succession of laminae; exterior of valve beyond the top edge of micrograph.
× 7000.

Fig. 3. View of internal surface of the secondary layer of a brachial valve of *Crania anomala* Müller showing the spiral growth patterns of laminae.
× 6000.

Fig. 4. Etched section of the secondary shell associated with an adductor scar in a brachial valve of *Crania anomala* Müller showing the development of a lense of prismatic calcite within a laminar succession; exterior of valve towards bottom left hand corner.
× 2600.

Fig. 5. Etched section of the secondary layer of a valve of *Craniops implicata* (Sowerby), Middle Silurian, Sweden, showing a succession of laminae; interior of valve beyond bottom edge of micrograph.
× 2500.

Fig. 6. Etched section of the secondary layer of a valve of *Trematobolus pristinus bicostatus* Goryansky, Lower Cambrian, Rassokha River Basin, U.S.S.R., showing laminae enclosing a lens of prismatic calcite; interior of valve beyond bottom edge of micrograph.
× 2500.

reached, a change occurs in the secretory habit of the outer epithelium. The calcitic crystallites are deposited as tabular rather than acicular aggregates and form a series of laminae usually about 250 to 300 nm thick and up to 15 μm across (Plate V, Figs 2, 3). This change in crystal growth is accompanied by the secretion of protein sheets, about 10 nm thick, which join up with one another to form an envelope for each lamina. The appearance of both the laminae and their organic covers marks the beginning of the secondary layer. On the internal surface of the layer, the laminae are seen as overlapping tablets of different sizes. Only rarely however, do they occur as perfect rhombohedra which had been increasing evenly in area in all directions at the moment of death. Mostly they had been growing spirally as single or double screw dislocations. The peripheral accretion arising from the perpetuation of a dislocation is usually shown on the surfaces of laminae as concentric banding with a periodicity of about 300 nm (Plate V, Fig. 3). If adjacent nuclei are crystallographically aligned, they may amalgamate. More commonly, junctions between contiguous laminae persist like fine cracks, while the edges of three or more non-aligned laminae may meet to define triangular or trapezoidal enclosures. Correlation in plan and section between the mineral skeleton and its associated decalcified tissue shows that protein sheets cover every surface of the spirally growing units except their edges, and that the plasmalemma of the secretory cell is stepped in complement to the different levels presented by adjacent laminae. The growth of this type of secondary shell is envisaged as follows.

A calcite seed is deposited either as an isolated rhomb or as the apex of a spiral growth feature already in existence. The seed almost invariably develops as a surface stepped by either or both left- and right-handed screw dislocations. Such steps ensure that further growth of the seed is by lateral accretion along edges of a developing spiral ramp. As the seed grows peripherally it simultaneously presents an expanding inner surface on which fabrication of a protein sheet can take place. But the lamina is normally part of a helicoid spiral so that its protein coat is also part of a continuous strip rotating around the dislocation line. Disposed in this way, the coat serves as a cover for one calcite surface and a foundation for the later-formed overlapping surface.

Secondary shell deposition in *Crania*, therefore, involves simultaneous secretion of protein and carbonate at different levels within a laminar succession. This mode of deposition is continuous not episodic (Williams, 1970). The distinction is important for palaeoecological research because a succession of laminae alternating with protein sheets does not constitute a reliable time series in the way that banded deposits do.

During secretion of both valves of *Crania*, branched outgrowths of the mantle acting as storage centres for polysaccharides and proteins, become enclosed in the primary and secondary shell, and although they differ structurally from the caeca of the Articulata the same terminology can be used to describe both types. The tubules forming the distal ends of a caecum in *Crania* are not organized like the terebratulide brush but splay out radially beneath the periostracum to which they are connected by only sporadically developed strands. Consequently, the punctae of *Crania* are immediately distinguishable from those found in living Terebratulida in the absence of distal sieve-like calcite canopies.

The skeletal succession just described is typical of the brachial valve except beneath the muscle bases where laminar deposition of secondary shell is subordinated to the secretion of either calcite blades disposed more or less vertically to the internal surface or lenses of prismatic calcite (Plate V, Fig. 4). Profound differences, however, have always been characteristic of the pedicle valve where normal growth seems generally to have been inhibited or grossly modified, presumably by the effects of cementation. In living *Crania* the pedicle valve is poorly developed compared with the brachial valve and only the periostracum and primary layer are secreted. In some extinct stocks like the Mesozoic *Isocrania*, on the other hand, a secondary layer composed of laminae and coarsely crystalline calcite is present. But the greatest variation is found among the earliest Craniacea of Ordovician age. In the unattached *Pseudocrania* both valves are alike in the full development of primary and laminar secondary shell. In contrast, the pedicle valve of the contemporaneous attached *Acanthocrania* is normally not preserved at all and was probably nothing more than a thickened periostracum for most of the life of the animal with only a thin impersistent arc of primary shell deposited in some individuals during late adult growth (Williams and Wright, 1970).

Despite the variability in the shell succession of fossil and living species, the standard secretory régime of the Craniacea may be regarded as involving six operations indicated by: an external mucopolysaccharide layer, an outer fibrillar triple-unit membrane, the main periostracal layer of mucopolysaccharide with fibrils, an inner protein sealing membrane, a primary calcareous layer of acicular crystallites, and a secondary calcareous layer of laminae encased in protein membranes (Fig. 2d). The succession is characteristic of the brachial valves of living species and, judging from fossil shells, has always been so. Both recrystallized primary and laminar secondary layers can be seen in the brachial valves of the earliest known genera and in the pedicle valves of two of them (the unattached *Pseudocrania* and the cemented *Petrocrania*) and it is reasonable to assume that the layers were deposited in those Ordovician stocks as they are in living *Crania*.

Two other inarticulate groups, both extinct, were equipped with calcareous shells. The Craniopsidae originated at about the same time as the Craniacea and their shells also are composed of a thin recrystallized primary layer and a laminar secondary layer (Plate V, Fig. 5) (Williams and Wright, 1970). However, shell morphology shows that the Craniopsidae in addition to being impunctate, were anatomically different from the Craniacea, and although there might be reservations about including them in the Lingulida, most students of the phylum do not regard them as being closely related to the Craniacea.

The Obolellida are known only from the Lower and Middle Cambrian, dying out long before the emergence of the Craniacea. Nonetheless, the ultrastructure of the obolellide shell, as seen in *Trematobolus*, is like the craniacean. A thin recrystallized primary layer is sharply distinguishable from a laminated secondary shell. The laminae tend to be lenticular in section, possibly reflecting some deformation during recrystallization, and are about 500 nm thick and up to 10 μm across. They are essentially flat-lying but become steeply inclined at regular intervals to accommodate nodules about 8 μm wide and high (Plate V, Fig. 6). The nodules may represent either primary patches of coarsely crystalline calcite secreted by outer epithelium, or secondary growths filling impersistent canals that originally contained short mantle papillae. Similar deposits either indicative of muscle platforms or filling punctae, are found in fossil Craniacea. Thus comparison between both groups is close enough to suggest that deposition of the obolellide shell was governed by essentially the same secretory régime as the shell of living *Crania* (Williams and Wright, 1970).

CONCLUSIONS

Much research remains to be done on the fabric of the brachiopod skeleton and fundamental changes in our views on the evolution of the structure of the calcareous shell may yet occur. In particular, the lack of information about the shell structure of Cambrian and early Ordovician brachiopods is the most critical gap in our knowledge. The deficiency arises from the rarity of specimens with adequately preserved shell fabric in rocks of such antiquity, but is all the more significant because phylogenetic differentiation of the shell was then taking place. Within that time, the secondary layer of the protein-calcitic shell of six orders became differentiated along two basic lines (Fig. 3) leading to either the fibrous layer of the Orthida, Pentamerida and Plectambonitacea, or the laminar layer of the Craniacea, Craniopsidae, Obolellida and Strophomenida (excluding the Plectambonitacea). Yet all morphological evidence unequivocally points to the laminar condition, at least, as having been polyphyletically derived. A primary layer essentially composed of acicular calcite crystallites must also have been the sole constituent of the shell of the Dictyonellidina, a small generalized

3. SEM Studies of Skeletons of Brachiopoda 61

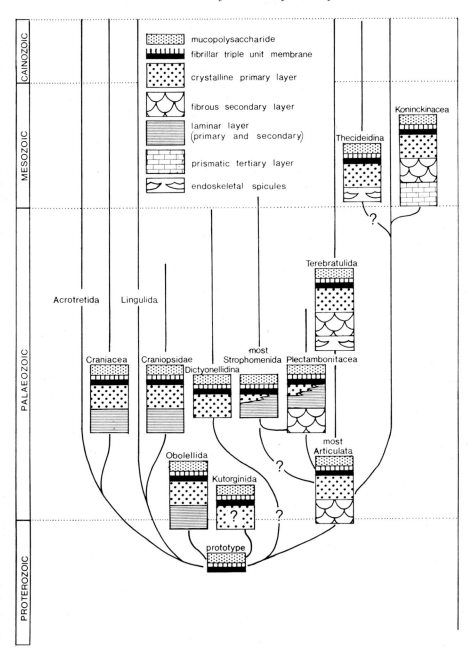

Fig. 3. Inferred phylogeny of the calcitic skeletal successions of the Brachiopoda.

group ranging from the Ordovician to the Permian and bearing no close resemblance to any other brachiopod group. In the absence of a reasonably complete Cambro-Ordovician record of brachiopod shell fabric, many conclusions recorded below can only be regarded as provisional.

The dominant shell fabric throughout the geological record of the Articulata involves a secondary layer of characteristically shaped and stacked fibres. It is exhibited by all orders, even the Strophomenida, and can be traced back to the earliest known Articulata the billingsellacean *Nisusia* and *Kotujella*. The presence of what appear to be punctae penetrating the shell of *Kotujella* is noteworthy because another orthide group, the Enteletacea, is also punctate although its earliest known species are Tremadocian in age and could not possibly have been derived from the early Cambrian *Kotujella* stock. On the contrary, shell morphology strongly suggests that the Enteletacea diverged directly from the Orthacea. In view of this and other evidence afforded by punctate stocks within the Rhynchonellida and Spiriferida, it is assumed that canals occupied by caecal outgrowths of the mantle developed repeatedly during brachiopod evolution. Whether all caeca penetrating the articulate shell were storage centres with essentially the same structure is unknown.

The consistent development of the fibrous shell in nearly all Orthida examined accords with the view that the order, as currently constituted, was the ancestral group of the Articulata. A fibrous secondary layer is also found in early Pentamerida like the Upper Cambrian porambonitacean *Bobinella*. The pattern is commonly modified in later Pentamerida by temporary cessations in the secretion of organic membranes so that fibres lose their individuality and are replaced by impersistent lenses of prismatic calcite. Further research may reveal that some pentameride species have a well-organized tertiary layer of prismatic calcite in the manner of some later Spiriferida. The periostracum of both the Orthida and Pentamerida is assumed to have been the simplest structure possible relative to those of living Articulata, and to have consisted of an outer mucopolysaccharide layer and an underlying fibrillar triple-unit membrane with a variably thickened base.

From these beginnings, the post-Cambrian evolution of the articulate shell structure is clearly defined. Descent of the Rhynchonellida from the Pentamerida probably did not involve any basic change in the skeletal succession except for the later development of a more elaborate periostracum. Derivation of the Spiriferida from the Rhynchonellida also probably occurred without any fundamental change in shell succession until after the order had become established. One of the changes was the gerontomorphic development of a tertiary layer of prismatic calcite in the Mesozoic Koninckinacea. In contrast, if the

Thecideidina evolved from the Spiriferida, a neotenous reduction of the calcareous succession also took place because, although a rudimentary fibrous secondary layer is found in some Mesozoic Thecideidina, only the primary layer persists in living species. Morphological comparison favours the descent of the Terebratulida from punctate Spiriferida and the groups are indistinguishable in shell structure. However, if the simple periostracum of the Thecideidina was also typical of the Spiriferida, the more elaborate periostracum of the Terebratulida would have been an important difference between exoskeletal successions.

Very little is known about the evolution of the endoskeleton. Spicules or plates, unconnected with the exoskeleton and evidently secreted within the connective tissue as in living species, have been recorded in Mesozoic species of the Terebratulida and Thecideidina. At present it is not possible to determine whether these traces represent a novel development restricted to the youngest orders emerging during brachiopod evolution, or whether similar accretions appeared sporadically in the mantle and lophophore of unrelated stocks throughout brachiopod history.

The laminar strophomenide shell is likely to have arisen in two stages. All Plectambonitacea which include the oldest strophomenide species known, have a secondary layer of normally developed fibres. Consequently, laminae could only have appeared in the primary layer of Plectambonitacea like *Bimuria* as paedomorphic replacements of the vertically disposed crystallites constituting the primary layer of all other contemporaneous Articulata. In the second stage of divergence from a typical articulate shell fabric, neotenous suppression of the fibrous secondary layer would then have given rise to the laminar shell succession of all other Strophomenida (including the Triplesiidina). It is questionable whether a vestigial primary layer composed of vertically disposed crystallites, was always secreted, as in other Articulata. Earlier work favoured the view that the layer was completely replaced by laminae. Current investigations, however, show that such a layer occurs in some stocks at least, either as a continuous external skin or as the proximal nodes of laminae forming overlapping concentric skirts ornamenting the shell exterior. Indeed a layer of crystallite aggregates may have always been deposited on the inner cementing surface of the periostracum precursory to the secretion of the laminar succession, although usually on so microscopic a scale as to be destroyed during diagenesis.

Irrespective of the degree of persistence of a normal primary layer, the laminar strophomenide shell was probably derived ultimately from the standard orthide shell. The morphology of the Plectambonitacea, the intermediate group in this line of descent, points to an orthide ancestry close to the Billingsellacea.

Such a link is strongly indicated by the presence of a pseudodeltidium and chilidium and the development of socket ridges in place of brachiophores. The morphology and shell fabric of early Billingsellacea, like *Kotujella* and *Nisusia*, are consistent with this link, although too far separated in time to have been the direct antecedents. The mid-Cambrian *Billingsella lindstromi* is temporally nearer the earliest Plectambonitacea but it has a laminar shell like that of Strophomenacea and Davidsoniacea. More work is now being done on the fabric of Cambrian shells, but it is already evident that, whereas the Plectambonitacea are unlikely to have diverged from the Billingsellidae, some Strophomenacea or Davidsoniacea might have done so, and that the Strophomenida might, therefore, be polyphyletic.

Whatever the relationships between the Strophomenida, Dictyonellidina and other Articulata, current stratigraphic as well as fabric studies support the opinion that the ancestral mineral skeleton of the Articulata consisted of a cryptocrystalline primary layer and a fibrous secondary layer. The associated organic covers were possibly an impersistent mucopolysaccharide layer and an underlying periostracum composed solely of a fibrillar triple-unit membrane. The fact that the primary layer appears to be significantly thinner in older groups previously prompted the idea that the mineral part of the protoypic shell would have consisted of only crude fibres sheathed in protein membranes and attached directly to the periostracum, and that deposition of the primary layer was a paedomorphic novelty introduced later into the articulate secretory régime. However, in view of the fact that a mineral deposit, structurally similar to the primary layer, is now known to underlie the periostracum of widely different phyla, it appears more likely that a zone of vertically disposed crystallites has always intervened between the periostracum and the fibrous layer of the articulate shell.

Other calcareous-shelled brachiopods can only be related to the Articulata through a soft-bodied prototype protected by a triple-unit fibrillar periostracum underlying an external coating of mucopolysaccharide. Attempts to find an identifiable shell structure in the Kutorginida have so far failed and comparison of their fabric with that of the articulate crystalline primary layer is provisional. In contrast, the shell of the Obolellida, Craniopsidae and Craniacea is known always to have consisted essentially of a primary layer with vertically disposed crystallites and a laminar secondary layer. There is, moreover, unequivocal evidence to show that, although the Obolellida may have arisen directly from a soft-bodied brachiopod prototype, the Craniopsidae and Craniacea were differently derived. Both these stocks appeared in the Ordovician long after the establishment of the phylum, and, for anatomical reasons, must be considered as having descended independently of each other from chitinophosphatic ancestors. The biochemical changes leading to the secretion of a protein-carbonate skeleton instead of a chitino-phosphatic one, are unknown at present. It is noteworthy that in the Brachiopoda at least, such a change appears to have been irreversible because no chitino-phosphatic stocks are known to have had protein-carbonate ancestors.

With reference to the ultrastructure of the calcitic exoskeleton, the laminar secondary shell is systematically less important than the monophyletic fibrous layer although it evolved on four different occasions during brachiopod history. It also seems to have been deposited in two fundamentally different ways. In the Craniacea and, judging from ultrastructural similarities, in the Craniopsidae and Obollida too, the laminar shell is, and was, built up spirally about dislocation lines with simultaneous secretion of protein and calcite on the faces and edges respectively of rhombohedral or dihexagonal plates. In the Strophomenida, on the other hand, the laminae consist of long narrow blades amalgamated with one another into relatively large sheets. Those belonging to one sheet must have been secreted simultaneously by groups of cells, and exudation of protein which is assumed to have separated successive sheets partially or completely from one another, might have alternated with calcite deposition. Neither process is closely comparable with the deposition of the standard secondary shell of the Articulata, which involves the simultaneous secretion of protein and calcite from the front and rear parts respectively of an epithelial cell. The precise nature of the laminae in the *Billingsella* shell has still to be determined, but it will be interesting to discover whether the laminae were originally plates or blades. That alone would greatly help in deciding the relationship between *Billingsella* and the Strophomenida.

REFERENCES

ARMSTRONG, J. (1969). The cross-bladed fabrics of the shells of *Terrakea solida* (Etheridge and Dun) and *Streptorhynchus pelicanensis* Fletcher. *Palaeontology* **12** (2), 310–320.

BAKER, P. G. (1970). The growth and shell microstructure of the thecideacean brachiopod *Moorellina granulosa* (Moore) from the Middle Jurassic of England. *Palaeontology* **13** (1), 76–99.

BIERNAT, G. and WILLIAMS, A. (1970). Ultrastructure of the protegulum of some inarticulate acrotretide brachiopods. *Palaeontology* **13**, 491–502.

BRUNTON, C. H. (1969). Electron microscope studies of growth margins of articulate brachiopods. *Z. Zellforsch.* **100**, 189–200.

JAANUSSON, V. (1966). Fossil brachiopods with probable aragonitic shell. *Geol. Fören. Förhandl.* **88**, 279–281.

JOPE, M. (1967). The protein of brachiopod shell. *Comp. Biochem. Physiol.* **20**, 593–605.

OWEN, G. and WILLIAMS, A. (1969). The caecum of articulate Brachiopoda. *Proc. R. Soc. B.* **172**, 187–201.

TOWE, K. M. and HARPER, C. W., Jr. (1966). Pholidostrophiid brachiopods: origin of the nacreous luster. *Science, N.Y.* **154**, 153–5.

WILLIAMS, A. (1968a). Evolution of the shell structure of articulate brachiopods. *Special Papers in Palaeontology* **2**. pp 55. Palaeontological Association, London.

WILLIAMS, A. (1968b). A history of skeletal secretion among articulate brachiopods. *Lethaia* **1**, 268–287.

WILLIAMS, A. (1970). Spiral growth of the laminar shell of the brachiopod *Crania*. *Calc. Tissue Research* **6**, 11–19.

WILLIAMS, A. and WRIGHT, A. D. (1970). Shell structure of the Craniacea and other calcareous inarticulate Brachiopoda. *Special Papers in Palaeontology* **7**, 51. Palaeontological Association, London.

4 | The Scanning Electron Microscope in Acarine Systematics

D. A. GRIFFITHS and J. G. SHEALS

*Ministry of Agriculture, Fisheries and Food,
Pest Infestation Laboratory, Slough, Buckinghamshire, England
and British Museum (Natural History), London, England*

"Nay in these moving atoms I could not onely see long bristles formerly specified but also the very hairs which grow out of their leggs, which leggs themselves are smaller than the smallest hair our naked eyes can discover." *Power, London,* **1664.**

Abstract: The value of scanning electron microscopy in acarine systematics is examined mainly in relation to mites classified in the orders Astigmata and Cryptostigmata. In the Astigmata the scanning electron microscope has facilitated the study in undistorted material of structures such as Grandjean's organ, the supra-coxal seta, the famulus and solenidia. The taxonomic significance of the variation observed in these structures is discussed. Small specific differences in the shapes of certain solenidia, which can be discerned only with great difficulty in mounted specimens, are seen clearly in scanning electron micrographs, and variations in the form of Grandjean's organ are shown to be important at generic level.

Amongst the Cryptostigmata, the scanning electron microscope has been used to examine a number of structures in mites classified in the families Phthiracaridae and Euphthiracaridae. Scanning electron micrographs are shown to be particularly useful for interpreting chaetotactic patterns on the notogaster and leg segments of these heavily sclerotized mites. On the leg segments variations in the relationships between setae and solenidia are illustrated, and in genera of the family Phthiracaridae a minute seta is shown to be coupled distally with one of the solenidia on the first tarsus.

INTRODUCTION

The mites or Acari comprise a large group of chelicerate arthropods which rival the insects in the extent to which they have colonized both aquatic and terrestrial habitats. They live in salt and freshwater, and the habitats of the terrestrial species extend from the lowest intertidal to the highest mountain zones; organic debris of all sorts can support vast populations. They are amongst the dominant animals in mineral soils, and in the organic soils of forests they may constitute

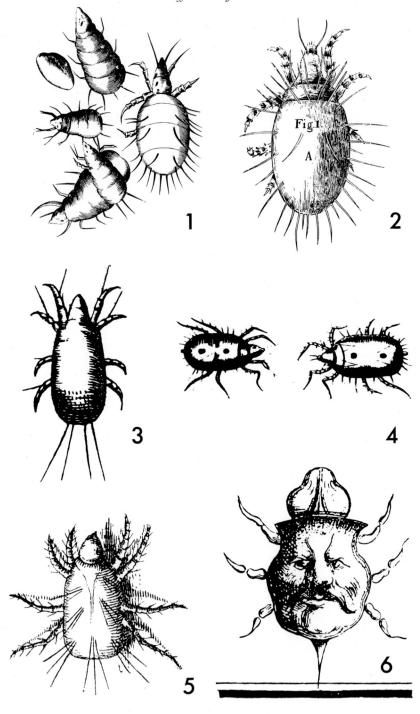

up to 7% of the total weight of the invertebrate fauna. Unlike other chelicerates, mites may be found living in close association with other animals, and these associations range from commensalism to obligatory parasitism involving both vertebrate and invertebrate hosts. The group has considerable economic significance. A number of phytophagous species are important as pests of garden and field crops. Other species infest stored food products and many parasitic forms are of medical and veterinary importance, either because of their activities as vectors of disease organisms or by virtue of their direct effects on the hosts.

In the classification scheme proposed by Evans *et al.* (1961), the Acari are divided into two major divisions, Actinochaeta and Anactinochaeta, comprising, respectively, three and four fairly well defined orders. Although some observations are included on other orders, this paper is concerned mainly with mites classified in two orders of the Actinochaeta, Astigmata and Cryptostigmata. One of us (D.A.G.) has been concerned mainly with the Astigmata and the other with the Cryptostigmata.

HISTORICAL

It is particularly appropriate that a symposium on a new technique of microscopy should include a contribution on mites for the history of acarology has always been very closely bound up with the history of microscopy. During the formative years of microscopy mites received a considerable amount of attention—Oudemans (1926–37), in his "Critical History", records over 3000 works published during the period 1650–1850 containing observations on mites—and this can probably be attributed to two factors, size and availability. The resolving power of the human eye at normal reading distances is about 100 μm, and the majority of mites fall into the 200–600 μm size range, that is to say large enough to be seen moving but too small to enable any detail of their structure to be observed with the naked eye. Animals in this size range would certainly have had a great appeal to early microscopists. Moreover, mites are amongst the most abundant and easily obtained animals. Any handful of forest leaf mould will

---

Fig. 1. "*Acarus casei*", from Griendl, 1687, *Micrographia nova* etc., Norimbergae.
Fig. 2. "Mite in cheese", from Hooke, 1665, *Micrographia*, London.
Fig. 3. "Mitte de la farine", from De Geer, 1778, *Memoires pour servir a l'histoire des insects*, Stockholm.
Fig. 4. *Acarus domesticus*, from Latreille, 1806, *Genera crustaceorum et insectorum*.
Fig. 5. *Acarus siro*, from Gessner, 1761, *Abh. naturf.*, Zürich.
Fig. 6. "Un animal nouveau", from Joblot, 1718, *Desc. microscopes avec observations d'insectes*, Paris.

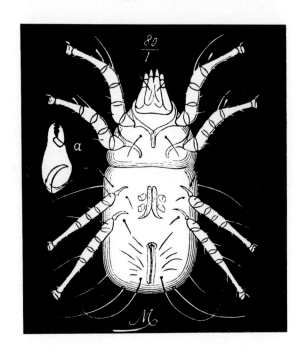

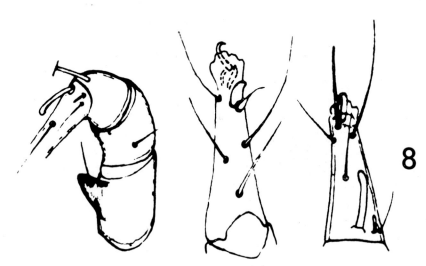

Fig. 7. *Acarus siro* L. (= *Tyroglyphus siro* Lat., *sensu* Mégnin, 1880) from Mégnin, 1880, *Les parasites et les maladies parasitaires*, Fig. 48.

Fig. 8. *Acarus siro* ♂, leg chaetotaxy, from Oudemans' original drawing.

almost certainly contain numbers of Cryptostigmata (as well as mites of other orders) and to obtain vast numbers of Astigmata the early microscopists could have needed to go no further than their larders to reach for the nearest cheese.

For the purposes of this symposium the development of acarine taxonomy during the period 1650–1850 needs only cursory attention. Although the later years of this period were characterized by a fever of activity, in general there was very little advancement. At the end of the period little was known of mites beyond the fact that they were (more often than not) eight-legged creatures and that their variously shaped bodies carried setae of various sorts. Figures 1–6, which include line drawings by such well known workers as De Geer and Latreille, serve to illustrate the work of this period, while Fig. 6 shows that on occasion the imagination of the observer could more than compensate for the poor resolving power of the microscopes of the period.

By the beginning of the second half of the 19th Century biologists were using microscopes fitted with Abbé's condenser and the first achromatic objectives. These improvements, together with the use of mounting media related to the refractive indexes of the lenses, resulted in the publication of the first really critical taxonomic works on mites based on comparative morphological studies. Mégnin (1880) for example, correctly mapped the distribution of the body setae of several Astigmata, including *Acarus siro* L. (Fig. 7) and towards the end of the 19th Century workers such as Oudemans, Michael, Berlese and Canestrini were describing leg chaetotaxy in some detail (see, for example, Fig. 8). However, whilst these workers made some use of leg-chaetotactic characters in the separation of taxa, detailed comparative studies of leg chaetotaxy were not undertaken until well into the 20th Century and in this field the work of Grandjean (1935 *et seq.*) is outstanding. He used the polarizing microscope to distinguish between various types of setae and seta-like structures on the appendages of the Actinochaeta (Fig. 9), and the nomenclatural system which he devised is now used by virtually all acarologists.

The detailed study of chaetotaxy in mites was also greatly facilitated by the introduction of phase-contrast microscopy. This technique is now regarded as an essential requirement for taxonomic work and it is particularly helpful for the study of the chaetotaxy of the idiosoma and leg segments in small lightly sclerotized species (see, for example, Figs 10 and 11).

Thus, in acarine taxonomy at the present time particular emphasis is laid on the nature of setae and seta-like structures and on the patterns of their distribution. But this is not to say that features other than chaetotaxy are ignored. Indeed, although a few authors may claim a phylogenetic basis for their groupings, most recently proposed classificatory schemes are in reality based on considerations of overall similarity, and chaetotaxy is important largely because this category of features provides a very large set of variates. Inevitably, therefore, the value of scanning electron microscopy in acarine systematics must depend to a great extent on its usefulness in chaetotactic studies.

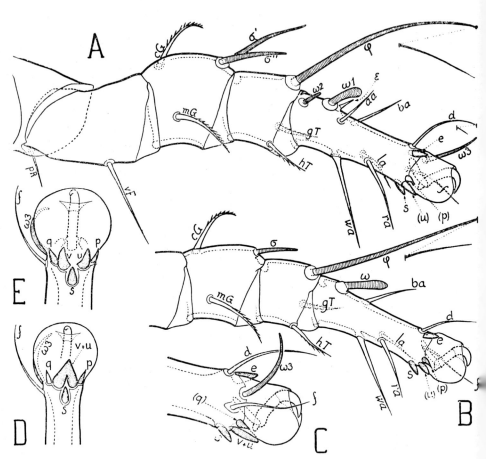

Fig. 9. *Forcellinia wasmanni* (Moniez), leg chaetotaxy and nomenclature, from Grandjean, 1939.

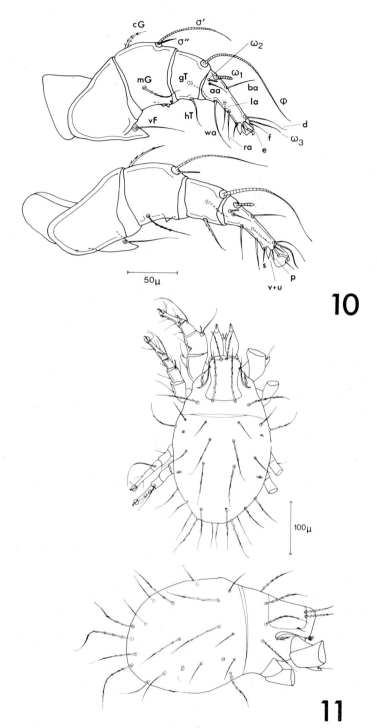

Fig. 10. *Acarus nidicolous* Griffiths, ♂. Idiosoma, dorsal and dorso-lateral views showing body chaetotaxy.

Fig. 11. *Acarus nidicolous* Griffiths, ♂. Leg I post-axial; top – British specimen, lower – unique specimen from U.S.A.

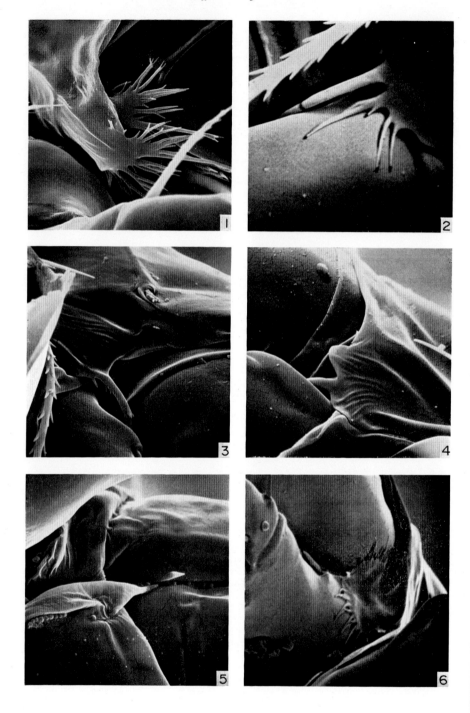

ASTIGMATA

The Astigmata are mainly weakly sclerotized mites which as adults range in size from about 200 to 1200 μm. The order includes a number of free-living species, several of which are economically important as pests of stored food products, as well as parasitic and epizooic forms. The majority of the parasitic forms are skin parasites of vertebrates.

Amongst the free-living Astigmata generic groupings tend to be natural, that is to say they are based on a consideration of many features, both morphological and biological. Until comparatively recently genera were separated by marked gaps, and while many of the features used in generic classification are difficult to observe, certain fairly prominent characters could be used as marks of the groups. However, with the very great increase in the number of described species, generic boundaries have become blurred and characters used formerly as markers no longer define natural groups. For example, at one time the *Rhizoglyphus* group of genera could be defined in terms of the relative lengths of the scapular setae, and before the discovery of *A. chaetoxysilos* Griffiths and *A. macrocoryne* Griffiths, species of the genus *Acarus* L. were characterized by their relatively short dorsal hysterosomal setae.

Thus, while all natural generic groupings can hardly be expected to be monothetic—that is to say definable in terms of single characters—there is a need in this group of mites for additional generic characters, and a number of structures which can be examined easily with the scanning electron microscope appears to be worth further study.

The first of these is a structure known as Grandjean's organ which is a cuticular protrusion situated on the lateral anterior margin of the propodosoma.

Plate I

Figs 1–6. Astigmata: Grandjean's organ (situated on the latero-anterior margin of the propodosoma).

Fig. 1. *Acarus farris* (Ouds.).
　×1100.
Fig. 2. *Tyrophagus putrescentiae* (Schrank).
　×1200.
Fig. 3. *Tyrolichus casei* Ouds.
　×1300.
Fig. 4. *Calvolia n. sp.*
　×1700.
Fig. 5. *Carpoglyphus lactis* (L.)
　×1920.
Fig. 6. *Suidasia* sp.
　×1920.

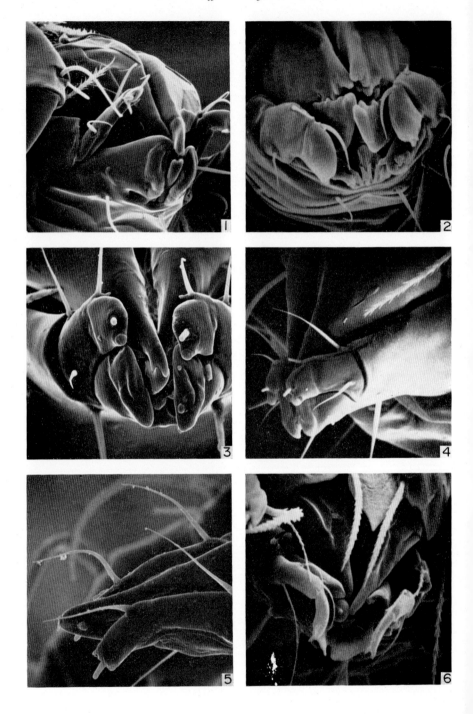

Because of its lateral position it is displaced during slide preparation and in mounted specimens it is generally hidden by the coxa of the first leg or by the surrounding coxal apodemes which are rather heavily sclerotized. This structure is very easily observed in undistorted material with the scanning electron microscope and its usefulness as a generic character is illustrated in the series of micrographs reproduced in Plate I.

Because of problems of distortion and the difficulties of interpreting three-dimensional structures in flattened preparations, the features of the gnathosoma in the Astigmata have received comparatively little attention. Preliminary observations with the scanning electron microscope suggest that the shape of the chelicerae, the form and setation of the palp and the form of the ventral lobes or malae could also be useful characters at generic level (Plate II).

Although in recent years some experimental taxonomic work has been carried out with free-living Astigmata, the majority of species have been erected solely on morphological criteria, and, as very few detailed studies of ontogeny have been undertaken, most species are defined only in terms of adult features. Moreover, in some genera it is not always possible to assign positively to species populations made up entirely of females, for species differentiation is often based on the form of the intromittent organ of the male and on male secondary sexual characters.

Criteria which have long been used for species separation include the relative lengths of certain setae on the dorsum of the hysterosoma, in particular those in the series d_1–d_4 (Fig. 12). However, the results of experimental studies have shown that, at least within the genus *Acarus*, these features are unreliable (Griffiths, 1964), and more recently it has been demonstrated that differences in the lengths of dorsal setae can be induced by culturing mites on diets of different

---

Plate II

Figs 1–6. Astigmata: gnathosomal features.
Fig. 1. *Glycyphagus domesticus* (De Geer).
  ×875.
Fig. 2. *Dermatophagoides farinae* Hughes.
  ×950.
Fig. 3. *Caloglyphus redikorzevi* (Zachvatkin).
  ×1740.
Fig. 4. *Tyrophagus nieswanderi* Johnston & Bruce.
  ×1000.
Fig. 5. *Carpoglyphus lactis* (L.).
  ×1300.
Fig. 6. *Gohieria fusca* (Ouds.).
  ×960.

nutritional values (Griffiths, 1970). Thus, the general value of this set of characters for species separation in the free-living Astigmata is very much open to question.

The form of a structure known as the supra-coxal seta has also been used for species differentiation, more often than not in conjunction with dorsal chaetotactic characters. This structure is borne on the hysterosoma slightly above the coxa of the first leg, and in mounted specimens it is frequently a problem to discriminate between "real" differences in its form and those due to distortion during preparation. Moreover, in flattened specimens the whole seta may be obscured by integumentary folds. It is extraordinarily easy to obtain good micrographs of this structure with the scanning electron microscope (Plate III) and the real significance of supra-coxal features, such as the number of pectinations and the lengths of the branches, which have been used for species differentiation, can now be easily investigated.

In addition to true setae, certain leg and palp segments in mites belonging to the Actinochaeta bear seta-like outgrowths which are known as *solenidia*—literally "little tubes". These structures were first recognized by Grandjean (1935) in cryptostigmatid mites. The solenidion is a thin chitinous tube, usually blunt ended and with a slightly expanded base, and the cuticular area from which it arises is generally slightly recessed. The solenidion is open at its proximal end so that its cavity is continuous with that of the appendage segment. The tarsus of the first leg of astigmatid mites usually bears three solenidia, and in the free-living genera, two (ω_1 and ω_2 in the terminology of Grandjean (1939)) are situated well within the proximal third of the segment, together with a structure known as the famulus. The latter, which is often conical in shape, is considered to be a modified seta since its walls contain birefringent chitin. In ectoparasitic and certain nest-dwelling Astigmata the famulus and all the solenidia are situated near the apex of the tarsal segment. A single solenidion (ω_1) is borne on tarsus II but tarsi III and IV have no solenidia.

The solenidion ω_1 is generally the longest and most prominent of the solenidia on tarsus I and its shape has been used for species differentiation by a number of workers, notably Zachvatkin (1941), Robertson (1959) and Hughes (1961). Often specific differences are quite marked (Plate IV, Figs 1, 2), but even so they can be difficult to observe since the specimen has to be oriented carefully so that the solenidion can be viewed laterally. Differences, however, can be very slight, for Griffiths (1964) showed that within the *Acarus siro* complex of species, the adult stages of two biologically well-defined species, *A. farris* (Oudemans) and *A. immobilis* Griffiths, were indistinguishable apart from very small differences in the shape of this solenidion on tarsi I and II (Plate IV, Figs 3, 4).

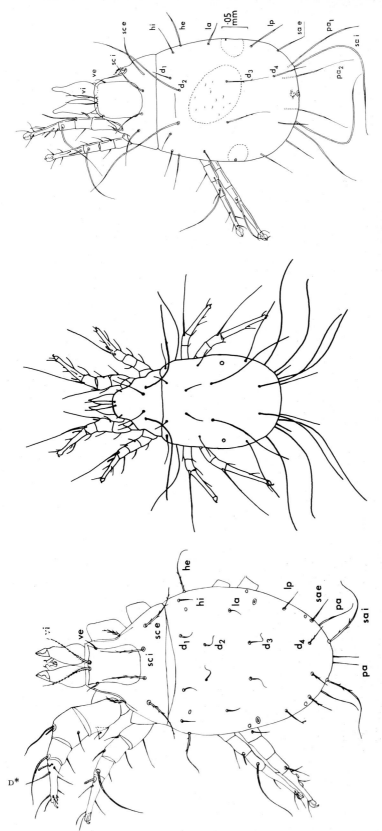

Fig. 12. *Acarus* spp. illustrating interspecific variation in the length of certain dorsal setae. Left to right; *A. siro* (L.), *A. tyrophagoides* Zachvatkin, *A. gracilis* Hughes.

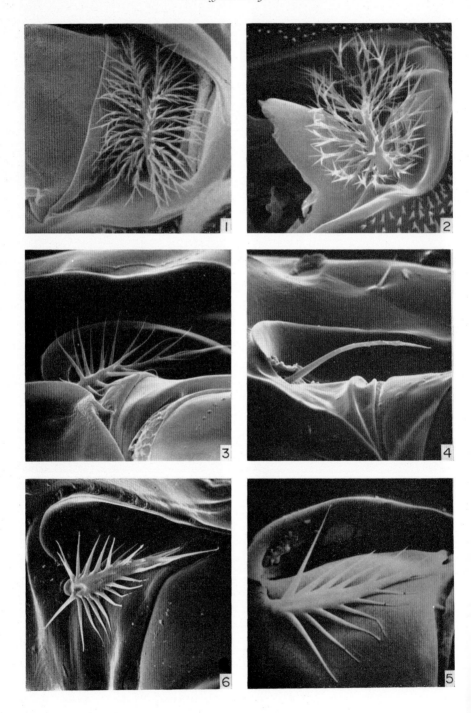

Solenidion ω_2 is located on the posterior surface of the first tarsus and is slightly proximal in relation to ω_1. The shape of this solenidion has rarely been used in species differentiation but scanning electron micrographs have revealed specific differences in some genera. These are well marked in a comparison of two species of *Glycyphagus* Hering, *G. destructor* (Schrank) and *G. domesticus* (De Geer) and a further study of variation in the form of this structure could well be worth while.

The form of the famulus has not been used as a taxonomic character in the Astigmata largely because this structure is very small and extremely difficult to study by conventional methods of microscopy. It is nearly always very closely associated with solenidion ω_1 and in slide preparations it is often obscured by this solenidion or by tarsal setae. Largely because material can be freely oriented under the probe it is very easy to obtain useful scanning electron micrographs of this structure (Plate IV, Figs 5, 6). It would appear that the famulus exists in a number of basic forms common to more than one genus and minor differences could provide features useful at specific level.

CRYPTOSTIGMATA

The Cryptostigmata, more commonly known as oribatid mites, are for the most part very heavily sclerotized species. All members of this order are free-living and they are especially abundant in the upper layers of forest soils. As adults they range in length from about 200–2000 μm, and whilst high magnifications are required for detailed studies of their morphology, the majority of species are too large and too opaque to be subjected to standard microscopical procedures. Notogastral features in large species are often studied with incident

---

Plate III

Figs 1–6. Astigmata: the supra-coxal seta.
Fig. 1. *Glycyphagus domesticus* (De Geer).
   × 2030.
Fig. 2. *Glycyphagus destructor* (Schrank).
   × 2050.
Fig. 3. *Caloglyphus redikorzevi* (Zachvatkin).
   × 1310.
Fig. 4. *Caloglyphus berlesei* (Michael).
   × 940.
Fig. 5. *Tyrophagus* n.sp.
   × 1750.
Fig. 6. *Tyrophagus perniciosus* Zachvatkin.
   × 1800.

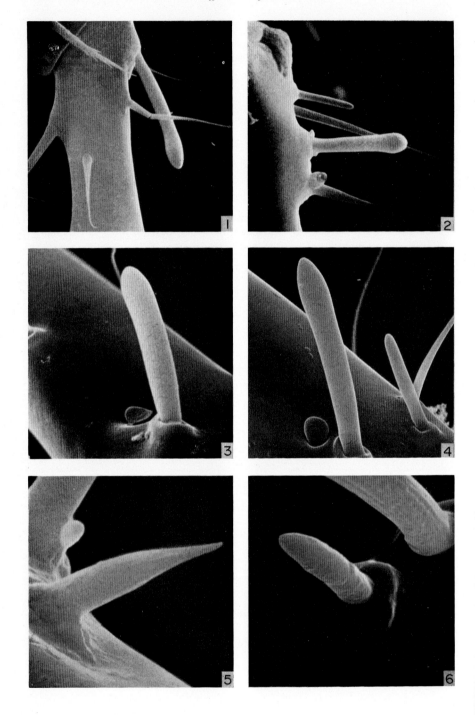

illumination using for example the moistened-carbon-block technique devised by Grandjean (1939), whilst small to medium sized species are generally examined with transmitted light in partially open cavity-slide preparations using lactic acid as a clearing and mounting medium. Although specimens can be oriented in cavity-slide preparations by using a fine needle or by manipulating the cover glass, for detailed studies of structures such as the gnathosoma and appendages specimens generally have to be dissected and their components mounted separately using either liquid media or gum chloral formulations.

The classification and identification of Cryptostigmata depends very substantially on the distribution patterns of setae on the notogaster, although features of the leg chaetotaxy are also useful at all taxonomic levels. The diversity of the leg chaetotaxy and solenidiotaxy and the patterns of their ontogenetic development have been studied in great detail by Grandjean (1935, 1939, 1961, 1967). These features have been shown to be well correlated with gnathosomal and idiosomal characters. Groups which can be defined in terms of the latter almost invariably have a distinctive arrangement of leg setae and solenidia, and within groups the pattern of ontogenetic development of these structures is also often characteristic.

Although the systematic value of leg chaetotactic characters is beyond dispute, these features are often neglected in descriptive work for their interpretation by ordinary methods of microscopy is often difficult and time consuming. For detailed study, legs have to be removed from the idiosoma—and with small heavily sclerotized species this can be extremely difficult—mounted and carefully

---

Plate IV

Figs 1–6. Astigmata: tarsus I sensory organs.

Fig. 1. *Caloglyphus redikorzevi* (Zachvatkin): solenidion ω_1 with a disc-like famulus at its base.
   ×1980.

Fig. 2. *Caloglyphus mycophagus* (Mégnin): solenidia ω_1, ω_2 and cone-like famulus.
   ×1500.

Fig. 3. *Acarus farris* (Ouds.): solenidion ω_1 with the famulus at its base.
   ×3100.

Fig. 4. *Acarus immobilis* Griffiths: solenidia ω_1, ω_2 and famulus.
   ×2750.

Fig. 5. *Carpoglyphus lactis* (L.): famulus.
   ×8000.

Fig. 6. *Tyrophagus nieswanderi* Johnston & Bruce: famulus.
   ×10,500.

oriented, and the origins of setae mapped out in optical section. Distortion during mounting can be a serious problem and the minute setae which are often closely associated with solenidia are sometimes very difficult to see.

A series of scanning electron micrographs are available of various structures in certain cryptostigmatid mites classified in a group which Grandjean (1967) has called the Euptyctima. In many respects these mites are amongst the most difficult Cryptostigmata to study and they provide excellent material for an evaluation of the usefulness of scanning electron microscope techniques. Known commonly as armadillo mites or box mites, the Euptyctima are an important component of the acarine fauna of soil and forest litter. The group, which is made up of four families, has presented a number of difficult classificatory problems and has recently been studied by numerical methods (Sheals, 1969).

Some of the main characteristics of the Euptyctima are illustrated in the micrograph of a whole specimen of a rather striking, and as yet unnamed, species of the genus *Steganacarus* Ewing (family Phthiracaridae) from forest litter in Israel (Plate V, Fig. 1). In this group of mites the legs can be retracted below the anterior part of the notogaster and completely covered over by the anterior shield or aspis which is hinged with the notogaster. The distributional pattern of the setae on the notogaster is a feature used in classification at all levels within the group. These setae are often very small and their distribution pattern on the notogaster (which may be as wide as it is deep) can be difficult to interpret. The curious pit located anteriorly on the dorsum of the notogaster of this mite (Plate V, Fig. 2) has not been observed in any other species although the dorsal cowl-like projection is known to occur in other phthiracarids.

Plate V

Figs 1–6. Cryptostigmata: *Steganacarus* spp.

Fig. 1. *Steganacarus* n.sp.: lateral aspect.
×45.

Fig. 2. *Steganacarus* n.sp.: dorsum.
×45.

Fig. 3. *Steganacarus* n.sp.: sensillus.
×340.

Fig. 4. *Steganacarus* n.sp.: infracapitulum.
×470.

Fig. 5. *Steganacarus striculus* (Koch): distal solenidion and associated sabre-like seta on tarsus I.
×5600.

Fig. 6. *S. striculus* (Koch): solenidion and associated seta on tibia IV.
×4250.

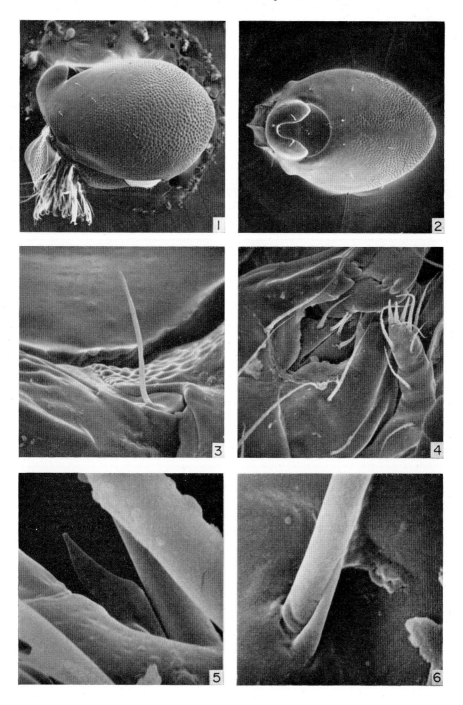

The chaetotactic and sculpture patterns on the dorsum of the aspis also provide useful taxonomic features, and the form of a trichobothrium known as the sensillus (Plate V, Fig. 3), which arises from the so-called pseudostigma near the posterior lateral margin of the aspis provides a useful character at species level. In some species the sensillus is expanded distally in two dimensions so that its appearance can depend very much on orientation.

The infracapitulum and chelicerae of the Euptyctima can generally be examined at high magnifications only in dissected specimens, and it is particularly interesting to be able to observe the undistorted structures (Plate V, Fig. 4). The heavily sclerotized rutella, which are often figured as flattened structures lying far apart, are shown lying close together with one chelicera in between them. The food-bearing chelicerae probably scrape against these structures as they move forward between them, and in this way food is particulated before dropping back above the rutella towards the mouth. The general form of the rutella shows little diversity within the Euptyctima, although in the so-called "higher" Cryptostigmata they are massive and completely obscure the lower lips. It is probable that the ability to feed on solid food particles is restricted to mites having rutella or rutella-like structures (cf. Plate II). Species without these structures must almost certainly feed on liquids or digest their food externally.

A number of features of the infracapitulum, particularly the chaetotaxy of the palp, are useful in classification, although in the Euptyctima major chaetotactic differences in this region are observed only at family level. It will be noted that in addition to the ordinary setae and the solenidion, the palp tarsus bears three cone-shaped setae of the type known as eupathidia. When viewed with transmitted light these conical setae are seen to be hollow, although unlike the solenidia their walls contain birefringent chitin.

Some of the features of the leg segments of the Euptyctima are particularly interesting and immensely useful in classification. For example, in certain species classified in the genus *Steganacarus*, notably *S. magnus* (Nicolet), the solenidion on the tibia of the fourth leg is well separated from a long anterior lateral seta. However, in other species currently classified in this genus, for example *S. striculus* (Koch), the solenidion on tibia IV is very closely associated with a small proximal coupling seta and the long seta in the anterior lateral position on this segment is absent. This coupling seta can be observed with the light microscope but its form and close association with the solenidion is seen much more clearly in scanning electron micrographs (Plate V, Fig. 6). The general pattern of setal arrangements in related mites provides fairly convincing evidence that the small coupling seta in *S. striculus* is the homologue of the long lateral seta of *S. magnus*. In a numerical taxonomic study of this group it was found that within the *Steganacarus* complex of species the mites with a coupled solenidion on the fourth tibia formed a fairly tight group and of course in this study groupings were based on considerations of overall similarity. Here differences in setal arrangements on tibia IV are consistently associated with other differences, notably differences in notogastral chaetotaxy which may be difficult to observe, and the presence or absence of a long seta in the lateral position on tibia IV provides a convenient marker character for the two *Steganacarus* groups.

The close association of solenidia and setae also occurs on other leg segments. For example on the tibiae of the first three legs of the Euptyctima the solenidia are invariably coupled. The lengths of the coupling setae and the closeness of the association vary. In one family, the Oribotritiidae, the coupling setae tend to be very long, often longer than the solenidia they accompany, while in another family, the Phthiracaridae, they are invariably very short—often barely discernible—and in this family they are located in the same depression of the cuticle as the solenidion. In the family Euphthiracaridae an intermediate condition is found. Here all the coupling setae are moderately long and although distally they are very closely applied to the solenidia their bases are quite distinct.

Scanning electron micrographs have also confirmed the suspected presence of a setal/solenidion association on the first tarsus of mites classified in the family Phthiracaridae. In *Steganacarus striculus*, for example, the distal solenidion on the first tarsus is coupled distally with a minute, delicate sabre-shaped seta (Plate V, Fig. 5). A similar association is found in *Steganacarus magnus* although the form of the coupling seta is rather different. This seta, which may be the homologue of the long seta coupled distally with the distal solenidion on tarsus I in certain Oribotritiidae, has now been found to occur in all genera of the Phthiracaridae. In all euphthiracarid genera, on the other hand, the distal solenidion on tarsus I is invariably free.

Associations of solenidia and setae have apparently evolved quite independently in several mite groups. Amongst the Prostigmata so-called duplex setae are found, for example, on the tarsi of tetranychoid mites. In *Tetranychus* species, including *T. urticae* Koch—the familiar glasshouse red spider mite, the coupling seta is relatively long and (in comparison with the Euptyctima) quite far removed from its solenidion. A similar association is found in *Bryobia* spp. It is interesting to speculate on the function of this association. Solenidia are almost certainly chemosensory (Grandjean, 1961), while normal setae are tactile, and the coupling arrangement could enable the solenidion to function as both a chemosensory and tactile organ. Purely tactile stimuli could be transmitted to the coupling seta at its base.

In three families of the Euptyctima the ano-genital region is very narrow and covered by a series of contiguous elongated plates which show varying degrees of fusion. Because of the great depth of the notogaster it is impossible, without dissecting specimens, to examine these ventral structures in detail with ordinary methods of microscopy. The detail of the system of interlocking lobes on the anal valves of the Euphthiracaridae is well shown in scanning electron micrographs (Plate VI, Fig. 2). In species of *Microtritia* and *Rhysotritia* these are confined to an anterior median triangle flanking the posterior border of the genital shield, but in species classified in the genus *Euphthiracarus* a system of interlocking lobes is also found on the posterior part of the anal valve.

Another interesting feature of euphthiracarid mites is the structure of their egg chorion. Very few mites have heavily ornamented eggs but in two euphthiracarid genera, *Rhysotritia* and *Microtritia*, the chorion when viewed with transmitted light appears to be very heavily sculptured with a tessellated pattern. A number of authors have commented on the very heavily sculptured eggs of *Rhysotritia* species and their gross features have often been figured (see for example Jacot, 1930) but the structure of the chorion in euphthiracarids was first studied in detail by Märkel and Meyer (1959). These authors published a series of conventional micrographs of the egg chorion of certain *Euphthiracarus* and *Rhysotritia* species including that of *Rhysotritia minima* (Berlese), a species now classified in the genus *Microtritia*. They observed that the chorion in *Rhysotritia* species was markedly tessellated, each plate having a central raised porose area and a series of tortuous marginal tuberosities. Adjoining plates were separated by a sinuous fusion line. In the case of *R. ardua* and *R. duplicata* they observed

---

Plate VI

Figs 1–6. Acari: Cryptostigmata (Figs 1–2), Prostigmata (Figs 3–4) and Mesostigmata (Figs 5–6).

Fig. 1. *Rhysotritia duplicata* (Grandjean): egg.
　　×490.

Fig. 2 *Euphthiracarus cribrarius* (Berlese): anterior interdigitating processes on anal valves.
　　×700.

Fig. 3. *Tetranychus cinnabarinus* (Boisduval): cuticular lobes.
　　×7700.

Fig. 4. *Tetranychus urticae* Koch: cuticular lobes.
　　×7870.

Fig. 5. *Dermanyssus gallinae* (De Geer): stigma.
　　×11,000.

Fig. 6. *Geholaspis longispinosus* (Kramer): peritrematal region.
　　×150.

4. The SEM in Acarine Systematics 89

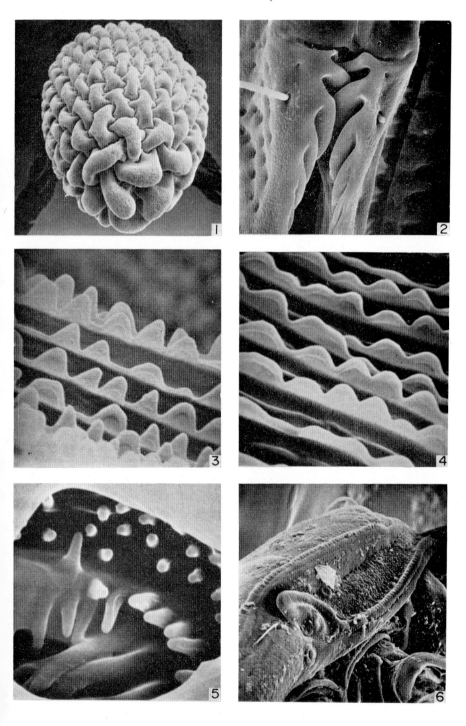

that the plates were arranged with their long axes parallel to the long axis of the egg, but in the case of *M. minima* the long axes of the plates were at right angles to the long axis of the egg. An ornamented chorion was also noted for *Euphthiracarus* species, but here the sculpturing was less marked and the pattern quite different.

A series of scanning electron micrographs of the eggs of *Rhysotritia* species has been obtained by dissecting spirit-preserved ovigerous females and allowing the eggs to air-dry on the stubs. In *R. duplicata* (Plate VI, Fig. 1) the chorion is made up of a series of porose interlocking segments, which in section can be seen to be traversed by numerous very fine canals. There is little evidence of fusion between the segments. In *R. ardua* and in an undescribed species of *Rhysotritia* the structure of the chorion is essentially similar and no specific differences have as yet been detected. With *Euphthiracarus* species this method of examining the eggs has not been successful. The chorion in these species is much more delicate and when air-dried the eggs shrink and contort badly. However, micrographs which have been obtained, although of poor quality, are sufficient to show that in species of this genus the sculpture pattern on the chorion is in the form of small rosettes and quite different from that found in *Rhysotritia* species.

OTHER ACARINE GROUPS

The Prostigmata, which is certainly the most heterogeneous of all the acarine orders, includes a number of phytophagous species of considerable economic importance, and in this order one of the more obvious applications of scanning electron microscopy is in the study of the lobe structure on the integumentary striae of red spider mites (family Tetranychidae).

In the genus *Tetranychus* Dufour the main criterion for species determination has been the shape of the male aedeagus, and *Tetranychus* populations made up entirely of females could rarely be assigned positively to species. However, in a study of North American material, Boudreaux (1956) found evidence of consistent morphological differences in both males and females which appeared to be useful for separating the components of the *T. urticae* group of species. For example, the lobes on the dorsal integumentary striae of *T. cinnabarinus* (Boisduval)—commonly known as the carmine mite—were reported to be semicircular or triangular, while in the common glasshouse red spider mite, *T. urticae*, they were said to be rather broader and larger and sometimes even oblong in shape. The general value of these characters has since been called into question, notably by van de Bund and Helle (1960) and by Robinson (1961). But Boudreaux and Dosse (1963) concluded that to some extent the lobes were useful characters in all *Tetranychus* species, and their illustrations of lobe variation in the genus included transmission electron micrographs of replicas showing very distinct differences in lobe structure between European material of *T. cinnabarinus* and *T. urticae* forma *dianthica* Dosse. They emphasize, however, that the appearance of the lobes will vary with the angle of observation and that a proper study of their form is impossible unless the striae are completely flattened against the cover glass.

With the scanning electron microscope the cuticular lobes are fairly easily seen in freeze-dried material, and each lobe is seen to have a thickened outer rim—a feature not easily seen in micrographs of replicas. Preliminary studies of British material of *T. urticae* and *T. cinnabarinus* have not revealed very marked differences in lobe structure. However, in *T. cinnabarinus* many lobes are distinctly triangular in outline (Plate VI, Fig. 3), while in *T. urticae* they are mostly semi-circular and tend to be rather broader (Plate VI, Fig. 4).

In the case of mites belonging to the order Mesostigmata, the scanning electron microscope has been found to be useful in the examination of a number of structures which are often difficult to see with the light microscope. For example, it is extremely easy to obtain micrographs of the deutosternal teeth on the infracapitulum and these structures provide useful taxonomic features in a number of groups. In the Mesostigmata tracheal trunks lead into a pair of stigmata situated laterally in the region of coxae II–IV, and associated with the stigmata are a pair of long anteriorly directed grooves of unknown function called peritremes. A number of interesting features of the peritrematal region have been revealed by the scanning electron microscope, although these have yet to be followed up in detail. Micrographs of *Dermanyssus gallinae* (De Geer), a blood-sucking parasite of birds, reveal the presence of numerous small papillae deep within the stigma and peritrematal groove (Plate VI, Fig. 5). Similar papillae are found in the stigma and peritreme of the free-living *Pergamasus* species, but in *Geholaspis* species, which are also free-living, it is interesting to find that the peritrematal groove and indeed the stigma itself is completely roofed over (Plate VI, Fig. 6).

SURVEY

The external morphological features of mites are extremely complex. Their bodies and appendage segments bear numerous setae and seta-like structures, which, more often than not, are arranged in very definite patterns, and while some of the larger species can be examined with a low-power binocular microscope and the proper relationships of structures observed, the vast majority have to be studied at high magnifications. Inherent in this method of examination are problems of distortion and the difficulty of gaining an impression of the true form of three-dimensional structures from studies in optical section.

Nearly all of the structures which have been described in this paper can be seen with the light microscope and the main advantage of scanning electron microscopy lies in the ease with which undistorted material can be manipulated under the probe. In this way the technique is immensely helpful for observing shape in structures such as solenidia and trichobothria, and for working out the distribution patterns of setae on the body and on the appendage segments.

Generally the most difficult structures to examine by conventional microscopy are those borne by the tarsus of the first leg. In many groups of the Actinochaeta this segment provides more useful taxonomic characters than any other leg segment, yet in the Anactinochaeta the comparative chaetotaxy of this segment has been virtually unstudied, largely because of the immense interpretive problem. A scanning-electron-microscope study of tarsus I in this group of mites, particularly the Mesostigmata, would almost certainly yield valuable results.

Quite obviously scanning electron microscopy has also considerable potential as an aid to the study of function in the Acari. Micrographs of the ano-genital and gnathosomal features, for example, are highly informative and more detailed comparative studies with this technique could help to reveal the purposes of structures, such as Grandjean's organ and the supra-coxal seta, which hitherto have only had classificatory significance. As Manton (1959) has pointed out, the functions which diagnostic characters serve concern the habits of life of the animals, and it is the elucidation of these habits which leads to an understanding of their evolution.

At a more routine level, the scanning electron microscope will be extremely useful in purely descriptive acarine taxonomy. At the present time the burden of descriptive work in acarology is rather heavy. About one thousand new mite species are described annually and this rate of description shows little sign of diminishing—indeed in some groups it appears to be increasing quite sharply. The wider use of scanning electron micrographs would undoubtedly improve the general usefulness of descriptions, and, apart from eliminating much of the subjective element in this sort of work, could do much to reduce the labour involved in compiling descriptive accounts.

While many completely new morphological features of mites will undoubtedly be discovered with the scanning electron microscope, its immediate potential in acarine systematics probably lies more with the proper interpretation of known features. The instrument will certainly be an invaluable aid to accurate description and its use will permit the accumulation of a vast amount of strictly comparable data for classificatory work. But while much of the data used by acarologists in the construction of natural systems may relate to structures which are difficult to examine, workers in systematic-dependent disciplines may derive some comfort from the knowledge that the marks of natural groups often prove to be features that can be discerned with a hand lens.

ACKNOWLEDGEMENTS

We are indebted to the staff of the Electron Microscope Unit of the British Museum (Natural History), particularly Mr B. S. Martin and Mr C. G. Ogden, for their advice and assistance. Figure 1 in Plate II is reproduced by permission of Cambridge Scientific Instruments Ltd.

REFERENCES

BOUDREAUX, H. B. (1956). Revision of the two-spotted spider mite (Acarina, Tetranychidae) complex, *Tetranychus telarius* (Linnaeus). *Ann. ent. Soc. Am.* **49**, 43–48.

BOUDREAUX, H. B. and DOSSE, G. (1963). The usefulness of new taxonomic characters in females of the genus *Tetranychus* Dufour (Acari: Tetranychidae). *Acarologia* **5**, 13–33.

BUND, C. F., VAN DE and HELLE, W. (1960). Investigations on the *Tetranychus urticae* complex in N.W. Europe (Acari: Tetranychidae). *Entomologia exp. appl.* **3**, 142–156.

EVANS, G. O., SHEALS, J. G. and MACFARLANE, D. (1961). "The Terrestrial Acari of the British Isles. I. Introduction and Biology." 219 pp, British Museum (Natural History), London.

GRANDJEAN, F. (1935). Les poils et les organes sensitifs portés par les pattes et le palpe chez les oribates. *Bull. Soc. zool. Fr.* **60**, 6–39.

GRANDJEAN, F. (1939). La chaetotaxie des pattes chez les Acaridiae. *Bull. Soc. zool. Fr.* **64**, 50–60.

GRANDJEAN, F. (1961). Nouvelles observations sur les oribates (1re série). *Acarologia* **3**, 206–231.

GRANDJEAN, F. (1967). Nouvelles observations sur les oribates (5e série). *Acarologia* **9**, 242–272.

GRIFFITHS, D. A. (1964). A revision of the genus *Acarus* L., 1758 (Acaridae, Acarina). *Bull. Br. Mus. nat. Hist. (Zool.)* **11**, 115–161.

GRIFFITHS, D. A. (1970). A further systematic study of the genus *Acarus* L., 1758 (Acaridae, Acarina), with a key to species. *Bull. Br. Mus. nat. Hist. (Zool.)* **19**, 85–118.

HUGHES, A. M. (1961). "The Mites of Stored Food." Ministry of Agriculture, Fisheries and Food Technical Bulletin No. 9. 287 pp, H.M.S.O., London.

JACOT, A. P. (1930). Oribatid mites of the subfamily Phthiracarinae of the northeastern United States. *Proc. Boston Soc. nat. Hist.* **39**, 209–261.

MANTON, S. M. (1959). Functional morphology and taxonomic problems of the Arthropoda. *In* "Function and Taxonomic Importance" (CAIN, A. J., ed.), pp. 23–32. Systematics Association Publication No. 3, London.

MÄRKEL, K. and MEYER, I. (1959). Zur Systematik der deutschen Euphthiracarini. *Zool. Anz.* **163**, 327–342.

MÉGNIN, P. (1880). "Les Parasites et les Maladies Parasitaires." 478 pp with volume of 26 figures. G. Masson, Paris.

OUDEMANS, A. C. (1926). Kritisch hirstorisch overzicht der acarologie. Eerste gedeelte. *Tijdschr. Ent.* **69** (Suppl.), 1–50.

OUDEMANS, A. C. (1929). Kritisch historisch overzicht der acarologie. Tweede gedeelte. *Tijdschr. Ent.* **72** (Suppl.), 1–1097.

OUDEMANS, A. C. (1936–37). "Kritisch Historisch Overzicht der Acarologie." Derde Gedeelte (6 volumes). E. J. Brill, Leiden.

ROBERTSON, P. L. (1959). A revision of the genus *Tyrophagus* with a discussion on its taxonomic position in the Acarina. *Aust. J. Zool.* **7**, 146–181.

ROBINSON, D. M. (1961). A species of *Tetranychus* Dufour (Acarina) from Uganda. *Nature, Lond.* **189**, 857–858.

SHEALS, J. G. (1969). Computers in acarine taxonomy. *Acarologia* **11**, 376–394.

ZACHVATKIN, A. A. (1941). [Tyroglyphoidea (Acari)]. *Fauna SSSR*, No. **28**, 1–475. (In Russian, English translation by RATCLIFFE, A. and HUGHES, A. M., American Institute of Biological Sciences, Washington, D.C.)

5 | The Reaction of Systematics to the Revolution in Micropalaeontology

P. C. SYLVESTER-BRADLEY
The University of Leicester, Leicester, England

Abstract: In micropalaeontology there has not only been an improvement of SEM techniques; there has also been the collateral development of the Projection X-ray Microscope. The vast increase of data made available demands revolutionary publishing methods. Micromorphological discoveries have given impetus to functional interpretation. The greater insight into ultrastructure is a stimulant to research into the physiology of shell secretion.

INTRODUCTION

The application of scanning electron microscopy to micropalaeontology has revolutionized existing concepts by suddenly increasing by many times the amount of morphological data available. A three-fold reaction is now taking place. First, the new techniques are being elaborated yet further. Secondly, new methods of publishing the results are being explored. Thirdly, the new data are being used to investigate the ultrastructure of the shell and the functional morphology of newly revealed characters.

ADVANCES IN TECHNIQUE

No one who has operated a scanning electron microscope is satisfied with the results. At first the micrographs may seem so beautiful that no improvement would seem possible. But soon dissatisfaction creeps in. We find that some other laboratory has produced better pictures than we have done ourselves. We find there is something in a micrograph that we just cannot see well enough; if only the resolution were a little better a problem could be solved.

So all of us are seeking improvements. Partly these depend on the ready cooperation of the manufacturers. Partly, they concern operational techniques.

1. Cleaning Techniques

In micropalaeontology, the cleaning of the fossil material is an all-important preliminary to microscopy. But all cleaning methods are attended with the problem that they may introduce minute artifacts. When palaeontologists study Recent material, they are faced with an additional problem. Recent shells frequently incorporate organic material that would not survive fossilization. It is not always easy to detect on a micrograph of a coated specimen which features of the shell are mineralized, and which are composed of unmineralized organic material. In ostracods, for example, hairs on Recent shells may be confused with spines. The micropalaeontologist therefore needs to remove any organic integument from Recent material before he can adequately compare it with fossil material.

Palynologists and micropalaeontologists who deal with acid-insoluble fossils have an easier task than others. The acid treatment itself most often yields material that is perfectly clean. Micropalaeontologists who deal with calcareous material are in the worst case, for often cleaning agents tend to erode the surface of microscopic material. Calcareous material may be cleaned either mechanically or chemically. The larger microfossils (0·8 mm or over) can be cleaned in water with a brush or fine needle, and this is often the best way of dealing, for example, with Palaeozoic ostracods. Plate I (Figs 2 and 3) shows what can be achieved by a skilled manipulator. It is very easy to break or damage specimens in this way, and it is not applicable to smaller material. The most usual mechanical method for these is to use an ultrasonic cleaner. Unfortunately, however, even this method has disadvantages. In cleaning ostracods I have found that pore canals are often enlarged, and their fine structure may be altogether destroyed. In Plate II, Fig. 1 it will be seen that the sieve-plate in a specimen cleaned ultrasonically for only two seconds has entirely disappeared; more than that—there has been a degree of recrystallization. Clearly, then, ultrasonic cleaning must be used with circumspection.

Unfortunately, if the dirt is fixed firmly to the specimen, as it often may be in

---

Plate I

Sleia troglodytophila (female right valve), an ostracod from the Silurian (Wenlock Limestone) of Wren's Nest, Dudley, Worcestershire, before and after cleaning under a stereo-microscope with a fine needle. Even with the advent of the SEM, manual dexterity in preparing fossils is at a premium. Specimen collected and prepared by David J. Siveter, of the University of Leicester.
Fig. 1. The specimen before cleaning.
Fig. 2. Intermediate stage. It can now be seen that the specimen is cracked. Presumably this developed during diagenesis.
Fig. 3. Clean.

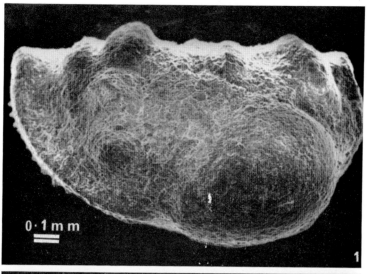

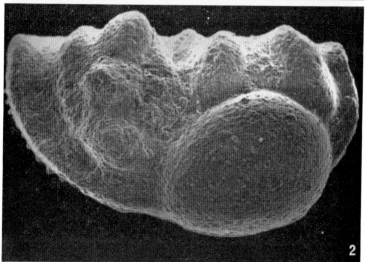

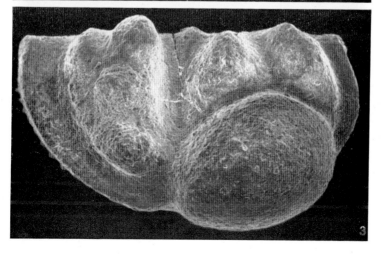

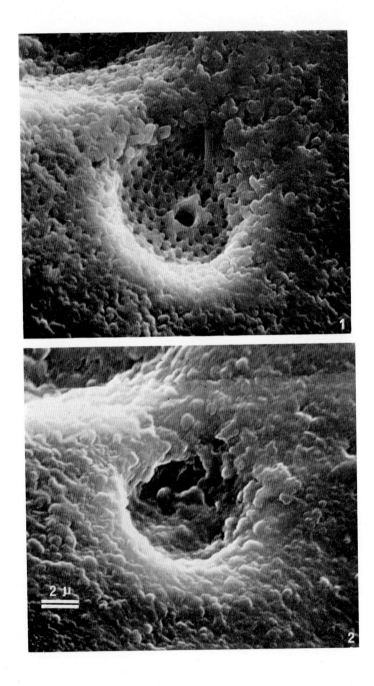

calcareous sediments, only mechanical cleaning can be effective. But in argillaceous sediments chemical cleaning can help. It is usual to use in the first place some mild alkali (e.g. sodium carbonate) or detergent. The most effective cleaning agent of this kind that I have tried is a commercial product, "Decon" (Medical-Pharmaceutical Developments Ltd., Brighton), intended for cleaning laboratory glassware. It is alkaline, and I use a 25% solution in demineralized water.

When cleaning Recent material, three methods may be effective in removing any organic integument. The most usual is by oxidation. Sodium hypochlorite or hydrogen peroxide may be used. In their SEM laboratory at Halifax, Nova Scotia, the Bedford Oceanographic Institute use a low temperature "asher" for this purpose. In this apparatus a stream of oxygen is excited by radio frequencies and organic substances can be ashed completely. Some care must be used, as the so-called "low temperature" may reach 200°C, or more, and in the presence of carbon this may calcine the delicate calcareous skeleton of a microfossil. Another effective method of removing organic matter is to steep the fossil for an hour or two in some form of protease.

2. Mounting Techniques

Really small microfossils (0·1 mm and below) may need no adhesive to keep them on the stub. A suspension in water evaporated to dryness is often all that is needed in mounting spores or microplankton, but larger fossils need an adhesive, and it is often useful to have one which allows one to manipulate a specimen so that it can be viewed in a special orientation. Many laboratories use double-sided cellotape for this purpose. I prefer a minute quantity of an adhesive which remains tacky, and so possesses some of the properties of cellotape. Two such adhesives (manufactured by photographic companies) are "Adhesyl" (Gevaert) and "Kodaflat" (Kodak).

3. Coating Techniques

Although a gold-palladium coating is the most often used in micropalaentology, it is not easily removed. If one needs to clean the coating off a specimen after it has been scanned, aluminium is to be preferred (Sylvester-Bradley, 1969).

Plate II

Fig. 1. The sieve-plate of a pore canal in the ostracod *Aurila convexa*.
Fig. 2. The same area of the shell after two seconds treatment in an ultrasonic cleaner. The fine structure of the sieve plate has been entirely lost by recrystallization.

This may easily be removed by immersing for a minute or two in a freshly made-up, dilute solution of alkali. Care must be taken not to let the solution evaporate, and thus get too strong, as delicate calcite spines can be eroded in concentrated alkali (Plate III, Figs 1, 2). I have found that the 25% solution of Decon used for cleaning effectively removes an aluminium coating.

Different coating materials often affect the contrast of the micrographs. This can be controlled to a certain extent by changing the operating conditions of the microscope, but often not without affecting the resolution also.

Some microfossils can be viewed without any form of coating. If a static charge does build up, it may sometimes be dissipated by squirting it with an antistatic spray (Plate IV, Figs 1, 2). This may be effective with either coated or uncoated specimens.

4. Increasing Resolution

All the manufacturing companies in the field are continuously improving their apparatus and this is often expressed in increased resolution. The best resolution in most instruments is however only obtained at higher magnifications. At low magnifications (which are often the more important for micropalaeontology) the best resolution obtainable is often much below that advertised for the instrument. To some extent this is dependent on the skill of the operator. Nevertheless, greatly increased resolution can be obtained by building up a montage at a greater magnification than required. This can then be photographically reduced (Plate V, Figs 1, 2).

5. Three-dimensional Effects

The ability to produce stereo-pairs (depending on the great depth of focus obtainable with scanning microscopes) is well-known, but sadly neglected. In my view, the additional information that can be provided by a stereo-pair is far in excess of that obtainable by either providing 2 pictures from different viewpoints, or by increasing the linear magnification by $\times 1\cdot 4$ (Plate VI, Fig. 1). The average interocular distance of the human is usually quoted as 65 mm, and this is often recommended as the best separation for stereo-pairs. Most viewers with

---

Plate III

Fig. 1. Misshapen spines in the ostracod *Bythoceratina scaberrima*.
Fig. 2. The same area after treatment in caustic soda. The solution was too concentrated and the spines have been etched.

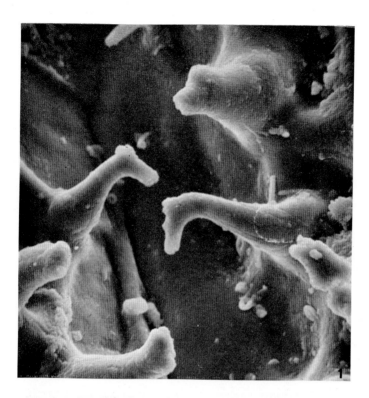

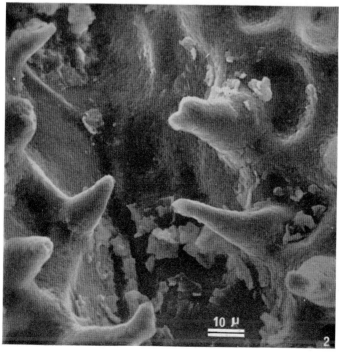

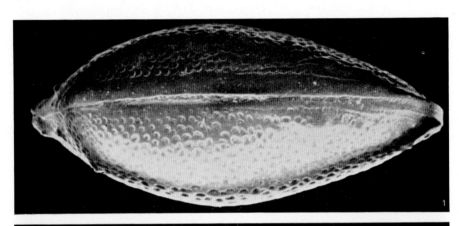
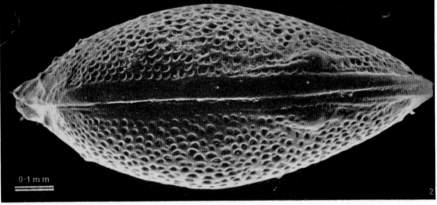

an interocular distance smaller than the average will, however, have difficulty at this setting, and 55 mm is much to be preferred. For a fairly flat object (hemispherical), a 10° tilt will give a satisfactory stereo representation; but for an object in which the depth is greater than the average diameter (for example, an ostracod seen in end view), a 5° tilt is better. It must be remembered that there is no "correct" angle, for this depends on the viewing distance that the stereo-pair is simulating. The two eyes of a person with an interocular distance of 65 mm will subtend an angle of 10° when viewing a point 37 cm distant which is close to the normal reading distance of most people.

6. Stereo Transmission Micrographs

Only a few years have passed since the commercial production of scanning electron-microscopes revolutionized micropalaeontology but already there has been another great leap forward. This has been signalized by the beautifully illustrated paper on planktonic foraminifera by Bé et al. (1969). These illustrations are the products of the Projection X-ray Microscope developed in the laboratories of the Technische Physische Dienst (Institute of Applied Physics at Delft).

The scanning electron microscope is supreme at revealing surface topography, but it is unable to give an indication of internal structures that lie below the surface. The pores and canals that penetrate the shell material remain hidden, but they are revealed with great brilliance by the Projection X-ray microscope. Moreover, the micrographs display the same great depth of focus that characterizes scanning electron microscopy and this means that stero-pairs can be prepared with ease.

The two techniques (SEM and PXM) are not in competition; they complement each other. Unfortunately the PXM has some limitations. The resolution obtainable is limited to about 2000Å; the SEM can achieve resolutions more than ten times as fine as that. Magnification in the PXM above about ×150 is therefore empty. This automatically limits the usefulness of the method to microfossils above about 0·5 mm in longest dimension. It is of particular interest, therefore, to workers in the Foraminifera and the Ostracoda, and in these two groups it is difficult to overrate its importance, for internal characters are not only revealed with much greater clarity than by any other method, but the

---

Plate IV

Fig. 1. *Aurila sp.* showing static charge although coated with aluminium.
Fig. 2. Same specimen as in Fig. 1 after treatment with anti-static spray.

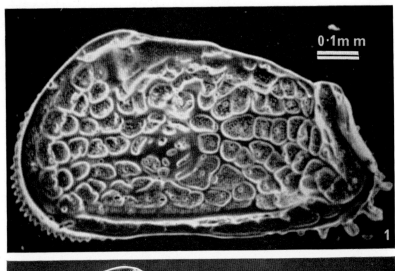

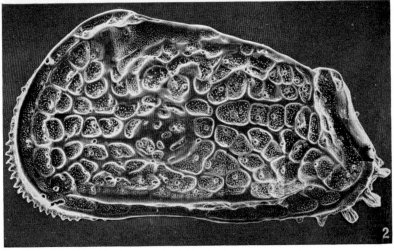

ability to view them stereoscopically enables one to trace the course of canals and other internal features that traverse the thickness of the shell (Bé *et al.*, 1969; and see Plate VII).

A specimen illustrated by both SEM and PXM stereo-pairs must reveal much more than has ever been seen before.

I understand that it would not be impossible to devise an adaption of the SEM which would enable one to use the instrument also for projection X-ray microscopy. The cost of a custom-built projection X-ray microscope is about two-thirds that of a production-model SEM.

PROBLEMS OF PUBLICATION

The beauty of the pictures provided by the new techniques gives such aesthetic satisfaction that all who work with them find themselves back in the days of the beginning of the "old systematics". We are faced with the same problems that faced the first microscopists. Our instruments are so expensive that we who have access to them are privileged. The objects we view are so intricate, so beautiful and so puzzling that we want to share them with as many of our colleagues as we can. The normal techniques of taxonomy afford us the best retrieval system in science, and modern techniques of lithographic printing can reproduce photographs so faithfully that we should have no difficulty in publishing our results. But here the very success of the instrument is a disadvantage. The SEM can in one day produce more micrographs worth publishing than most scientific journals are prepared to print in a whole year, for high-quality lithography is not cheap. Worse than that: if we were to produce new periodicals solely for the purpose of publishing SEM micrographs, we would find the field highly competitive, for scientific publications have proliferated at such a rate, and financial policies initiated by some commercial publishers have so enhanced costs, that normal scientific publication has almost reached saturation level. Somehow we must find a way of by-passing these difficulties. Professor Alwyn Williams has suggested during this symposium that what we need is a system whereby the individual scientist can buy just the pictures he wants—not a

---

Plate V

Fig. 1. *Patagonacythere wyvillethomsoni*. From a single negative; resolution is mediocre.
Fig. 2. Reduced from a montage of many separate prints at an original magnification of ×1,000. There is considerable increase in resolution, the limiting factor in this case being the method of reproduction.

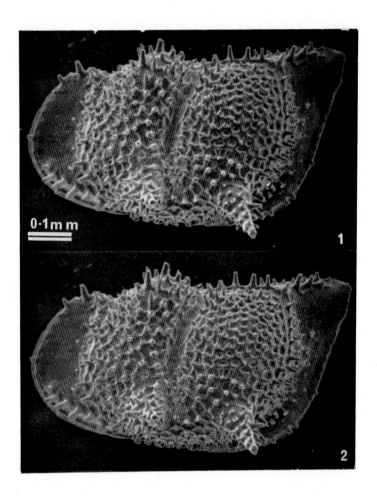

subscription to a journal of which the greater part is of no interest to him. We need to be able to buy the micrographs illustrating a single species. It is false economy to force us to purchase at the same time the pictures of nine other species that we do not want to see. Whether such a system of publication could ever work has still to be determined. I have myself been attempting to initiate the publication of an atlas of ostracod micrographs. I shall do my best to put Professor Williams's suggestion on trial.

One of the problems that bedevils publishing procedure in palaeontology, as it does in geology and in biological sciences generally, is the confusion that results from an attempted compromise. Martinsson (1969) has made this point when he contrasts the *idiographic* and *nomothetic* contents of a paper. In idiography we present data. A nomothetic paper uses the data in the development of theory. Too often an attempt at combining both aspects buries ideas under facts, and disperses factual data so widely as to hinder retrieval and collection. Most scientific journals are better designed to be nomothetic than idiographic. What we need in presenting the data provided by SEM is a method of publishing which is entirely devoted to the idiographic.

PROBLEMS OF INTERPRETATION

Of the reactions to the SEM revolution, one brings particular satisfaction. That is the result of the stimulus that has been given to research into functional morphology. Every new character revealed must have a meaning, and the search for function has spread into every field of micropalaeontology. This is not the place to elaborate detail. Two examples from my own area of specialization, the Ostracoda, will suffice to illustrate the sort of question that the SEM poses.

1. The Function of Hairs and Bristles

In the Ostracoda, bristles mounted on the test always seem to connect with pore-canals, whereas hairs seem merely to be extensions of the integument, though possibly sometimes mounted on papillae (Plate VII, Fig. 1). Presumably bristles are sensory whereas hairs have a mechanical effect (protection, insulation or frictional purchase).

Plate VI

Figs 1 and 2. A stereo-pair of *Bythoceratina scaberrima* (Brady); if viewed through a pocket stereoscope, a fully three-dimensional view is obtained.

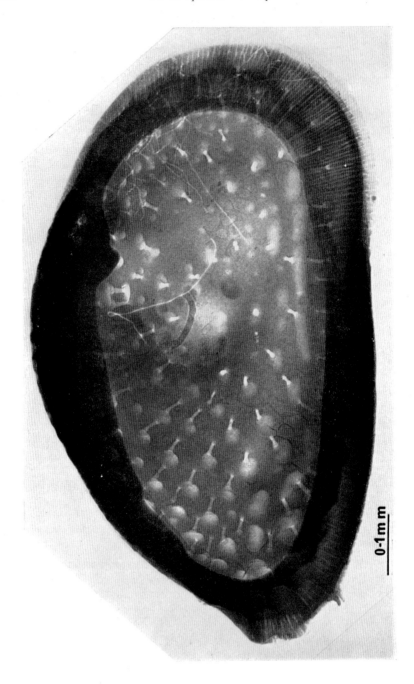

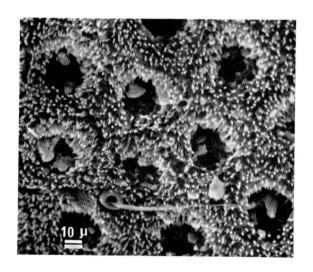

Plate VIII (above)

Hairs and a seta protruding from a pore canal in the freshwater ostracod *Ilyocypris nitida*.

◀ **Plate VII** (opposite)

Aurila convexa (Baird), a Projection X-ray micrograph showing course of pore canals through shell. Micrograph taken in the Department of Anatomy in the University of Amsterdam by courtesy of Dr. Booersma.

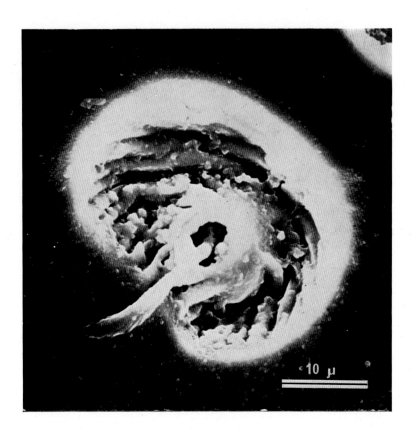

Plate IX

Bensonocythere whitei Swain. A seta protruding from a sieve-plate pore-canal which has been overgrown by an outer layer of the shell which is presumably secreted later than the sieve-plate.

2. The Physiology of Shell Secretion

In an ostracod, the shell of each instar appears to be secreted according to a time-sequence that establishes a succession in which the inner layers seem to be the older. In this way pore-canals may be earlier features which subsequently become "over grown" by material deposited on top of them (Plate IV, Fig. 1).

REFERENCES

Bé, A. W. H., Jongebloed, W. L. and McIntyre, A. (1969). X-ray microscopy of recent planktonic foraminifera, *J. Paleontology* **43**, 1384–1396.

Martinsson, A. (1969). Publishing in the geological sciences, *Lethaia* **2**, 73–86.

Sylvester-Bradley, P. C. (1969). Aluminium coating in scanning electron microscopy, *Micropaleontology* **15**, 366.

6 | Intérêt du Microscope Électronique à Balayage dans la Définition des Critères Génériques chez les Coccolithophoridées Fossiles

DENISE NOËL

Laboratoire de Géologie du Muséum National d'Histoire Naturelle, Paris

Abstract: In our present state of knowledge, the classification of fossil coccolithophorids is morphologically based. It is, therefore, very important for the micropaleontologist to describe with a great precision the different species found in the sediments.

With the scanning electron microscope it has become possible to study the coccoliths themselves instead of their carbon replicates as is necessary with a transmission electron microscope. It has also become possible to orientate these nannofossils at various angles. Thus use of this instrument allows a more direct and complete knowledge of this important group of fossils. Consequently the definition of generic criteria attains a greater precision. Some examples are chosen among upper cretaceous genera: *Ahmuellerella, Eiffelithus, Stradneria, Cylindralithus, Markalius*.

Moreover, because the SEM permits us to follow the constituent elements of the coccoliths from the distal to the proximal face, it has become possible to obtain a more accurate knowledge of the structure of these nannofossils and valuable information about relationships between the genera.

Résumé: Parce qu'il permet d'observer les coccolithes eux-mêmes et non leur moule en carbone, parce qu'il offre la possibilité de les orienter sous des angles variés, le microscope électronique à balayage nous donne une connaissance plus directe et plus complète des coccolithes fossiles. Par là, la définition des critères génériques atteint une plus grande précision. Des exemples sont choisis dans des genres crétacés.

De plus, précisément parce que l'étude des coccolithes en microscopie à balayage, permet de suivre les éléments constitutifs d'une face à l'autre du corpuscule, il devient possible d'acquérir une connaissance plus réelle de ces nannofossiles, en particulier en ce qui concerne leur structure.

INTRODUCTION

Fondée exclusivement sur la considération des coccolithes, seules portions de l'Algue originelle susceptibles de traverser les temps géologiques en se conservant, la classification des Coccolithophoridées fossiles est purement morphologique. Cela tient essentiellement au fait que les données biologiques sur les espèces vivantes sont encore bien fragmentaires et que les lignées phylétiques fort hypothétiques ne constituent pas des fils directeurs incontestables. De plus, parce qu'elles permettent d'observer les coccolithes avec beaucoup de précision, les méthodes modernes d'investigation—et la microscopie électronique en particulier—conduisent à attribuer une place prépondérante à ces critères morphologiques. Ajoutons encore que du fait de l'importance de ces nannofossiles en tant que constituants des roches carbonatées et en tant que marqueurs stratigraphiques démontrés ou éventuels, les études sur les coccolithes fossiles se sont multipliées au cours de la dernière décennie, nous livrant ainsi une masse importante de documents.

MICROSCOPIE ELECTRONIQUE A TRANSMISSION

Classification morphologique donc. Précisons qu'il n'existe pas—à ma connaissance—dans la nature actuelle deux espèces porteuses des mêmes coccolithes et qu'ainsi donc l'étude des corpuscules isolés donne une bonne idée de la diversité du groupe. Il existe cependant dans les mers et les océans des coques dimorphes revêtues en certains points de la cellule (ceinture équatoriale ou zone périflagellaire par exemple) de coccolithes de forme particulière, différente de celle des autres coccolithes du revêtement. Bien évidemment dans les espèces fossiles, un tel dimorphisme est impossible à reconnaître dès lors que l'on étudie les coccolithes isolés les uns des autres. La conséquence en est que le paléontologiste traite peut-être en tant qu'espèces distinctes—voire même en tant que genres distincts—les éléments isolés d'une même cellule. Il n'y a guère de possibilités d'éviter cet écueil.

C'est un truisme que de dire qu'une classification morphologique repose sur une bonne description des objets qu'elle ordonne. Cependant pour les coccolithes —et les nannofossiles calcaires en général—cette description n'a pas toujours pu être aussi précise que nécessaire, en raison de la petite taille de ces corpuscules dont les dimensions se comptent en microns. Il a fallu l'utilisation de plus en plus intensive de la microscopie électronique pour accéder à une bonne connaissance de la morphologie et de la structure des nannofossiles calcaires.

Le microscope électronique à transmission, parce qu'il ne permet d'observer que des objets transparents aux électrons a rendu indispensable l'utilisation de l'artifice que sont les répliques de carbone. Cependant l'examen d'une telle réplique de carbone ne donne du coccolithe qu'elle a moulé qu'une image statique: elle montre les éléments constitutifs du corpuscule avec une netteté remarquable mais ne permet pas—sans intégration des différentes images obtenues sur des coccolithes diversement orientés et appartenant avec quelque certitude à la même espèce—d'avoir une vue dans l'espace de ces éléments constitutifs les uns par rapport aux autres.

Nous perdons de la sorte un certain nombre de caractères tels par exemple la hauteur réelle,* du coccolithe étudié, l'appréciation exacte des concavités ou des protubérances que peuvent présenter les corpuscules, etc.

Rappelons cependant—malgré les réserves qui viennent d'être mentionnées—que la microscopie électronique à transmission a fait faire et continue de faire faire d'immenses progrès à nos connaissances sur les Coccolithophoridées.

MICROSCOPIE ELECTRONIQUE A BALAYAGE

Le microscope électronique à balayage, parce qu'il permet d'observer les objets eux-mêmes et non leur moule, parce qu'il nous offre la possibilité d'orienter et d'examiner ces objets sous des angles variés, nous conduit à une connaissance plus complète et plus précise des nannofossiles calcaires. Il est donc un outil précieux dans la recherche de caractères morphologiques pouvant servir de critères génériques et spécifiques.

Quelques exemples choisis parmi des Coccolithes du Crétacé supérieur (Craies sénonienne et maestrichtienne) illustreront cet intérêt du microscope électronique à balayage dans la définition de tels critères génériques.

Chez les coccolithes secondaires, il est possible d'individualiser un certain nombre de groupes morphologiques, assez nettement séparés les uns des autres, groupes auxquels on attribue généralement le rang de famille. La nomenclature que j'utiliserai dans cet article sera celle définie dans le mémoire suivant: D. Noël (1970a). "Coccolithes crétacés: la craie campanienne du Bassin de Paris".

La famille des Zygolithaceae Noël (1965) groupe des coccolithes elliptiques comportant une couronne marginale faite de lamelles de calcite inclinées sur les plans radiaux de l'ellipse dans laquelle s'inscrit le coccolithe. Cette couronne marginale limite une aire centrale diversement construite, quelquefois pourvue d'une hampe axiale, ou bien ne constitue qu'une sorte d'anneau dont les bords internes sont reliés par une structure transversale de construction variée.

Au sein de cette famille, la différenciation générique repose conventionnelle-

---

* Avec les répliques de carbone, "l'ombre" due à la métallisation effectuée sous un angle déterminé, donne dans certains cas une idée de la hauteur des coccolithes moulés. Mais ce n'est là qu'indication, impossible à traduire en une valeur exacte. Le hasard fait parfois que les corpuscules—en raison de leur forme ou parce qu'ils trouvent un support providentiel qui leur permet une position normalement instable—se présentent quelquefois de profil sur la préparation et sont donc moulés dans cette position. Mais dans ces cas là, si la hauteur et les caractères liés au profil deviennent observables, ce sont ceux liés à la vue apicale—de la face distale ou de la face proximale—qui ne le sont plus.

ment sur les caractères architecturaux et structuraux des différentes portions du coccolithe.

Par exemple, voyons pour le genre *Ahmuellerella* Reinhardt, 1964 (Planche I, Fig. 1a, b, c, d, e), les précisions morphologiques que nous livre l'examen au microscope électronique à balayage des coccolithes de ce type.

La figure 1a donnant, en ce qui concerne l'orientation—c'est à dire une vue apicale—du coccolithe, une image comparable à une réplique de carbone montre certains caractères utiles à la détermination du genre: forme générale elliptique; couronne marginale relativement étroite faite de lames de calcite se recouvrant largement les unes les autres; hampe axiale, traversée d'un canal circulaire, soutenue par huit travées rayonnantes qui ont valu dans le cas présent à ce coccolithe son épithète spécifique: *octoradiata*. Les Fig. 1 b, c, d, e montrent le même individu sous des angles différents et permettent de préciser la définition du genre: couronne marginale sensiblement aussi haute que large; hampe axiale dépassant très largement l'embase du coccolithe et soutenue par huit contreforts donnant à l'aire centrale une forme nettement conique. Tous ces caractères *deviennent donc mesurables* de façon précise sur le même individu.

De même pour le genre *Eiffelithus* (Reinhardt) Reinhardt (1966) (Planche I, Fig. 2a, b). L'examen en vue apicale—comparable du point de vue orientation du coccolithe à une réplique—nous montre (Planche I, Fig. 2a) la paroi de

---

Planche I

Fig. 1a, b, c, d: *Ahmuellerella octoradiata* (Gorka) Reinhardt. Le même individu observé sous des angles différents.
  × 5500.
  1a : Cliché S. 2966 $\alpha = 0°$
  1b : Cliché S. 2970 $\alpha = 60°$
  1c : Cliché S. 2971 $\alpha = 75°$
  1d : Cliché S. 2972 $\alpha = 85°$
  Maestrichtien supérieur de Dania (Danemark).
Fig. 2a, b: *Eiffelithus eximius* (Stover) Perch-Nielsen.
  × 5800.
  2a : Cliché S. 1639 $\alpha = 0°$
  2b : Cliché S. 1638 $\alpha = 45°$
  Campanien d'Arpenty (Essonne-France).
Fig. 3a, b, c: *Stradneria limbicrassa* Reinhardt.
  × 5000.
  3a : Cliché S. 1389 $\alpha = 0°$
  3b : Cliché S. 1388 $\alpha = 45°$
  3c : Cliché S. 1391 $\alpha = 72°$
  Campanien d'Arpenty (Essonne-France).

6. SEM des Coccolithophoridées Fossiles 117

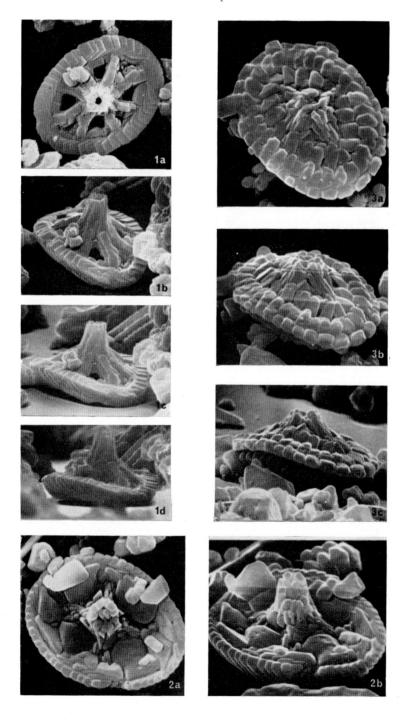

l'embase, extrêmement étroite, une structure centrale cruciforme, et entre cette structure centrale et la paroi, des cristaux d'assez grande taille. L'observation du même coccolithe de profil (Planche I, Fig. 2b) permet de préciser le rapport entre la largeur et la hauteur de la paroi de l'embase, montre que la structure centrale se développe en hampe axiale à partir du plancher du coccolithe et que les cristaux font saillie au-dessus de l'embase.

La famille des Podorhabdaceae Noël (1965) groupe des coccolithes dont la couronne marginale est faite d'éléments en disposition radiaire. Là encore, le microscope électronique à balayage permet d'affiner les descriptions et de définir des genres nettement délimités.

Le genre *Stradneria* Reinhardt (1964) illustré ici (Planche I, Fig. 3a, b, c) semble très voisin des genres *Polypodorhabdus* Noël (1965) et *Cretarhabdus* Bramlette et Martini, 1964 lorsque l'on examine ses représentants à l'aide de répliques de carbone. L'étude en microscopie à balayage permet de les différencier facilement en tenant compte des liens et des rapports entre les divers éléments constituant la couronne marginale (Noël, 1970a).

Autre exemple de description morphologique précisée par des examens en microscopie à balayage: *Markalius circumradiatus* (Planche II, Fig. 2a, b, c). La possibilité d'observer un tel coccolithe à la fois en vue apicale (Planche II, Fig. 2a) et de profil (Planche II, Fig. 2c) permet de définir très exactement la forme

---

Planche II

Fig. 1a, b, c: *Cylindralithus biarcus* Bukry (1969).
    ×5500.
    1a : Cliché S. 1695 $\alpha = 0°$
    1b : Cliché S. 1694 $\alpha = 45°$
    1c : Cliché S. 1702 $\alpha = 72°$
    Campanien d'Arpenty (Essonne-France).

Fig. 2a, b, c, d: *Markalius circumradiatus* (Stover) Perch-Nielsen (1968).
    ×3000.
    2a : Cliché S. 2767 $\alpha = 0°$
    2b : Cliché S. 2770 $\alpha = 30°$
    2c : Cliché S. 2772 $\alpha = 55°$
    2d : Cliché S. 2778 $\alpha = 82°$
    Maestrichtien inférieur de Møn (Danemark).

Fig. 3. *Gartnerago obliquus* (Reinhardt) Noël (1970b).
    ×3200.
    face proximale. Cliché S.2258; Coniacien d'Arpenty (Essonne-France).

Fig. 4. *Kamptnerius magnificus* Deflandre (1959).
    ×2500.
    face proximale. Cliché S.2395; Coniacien d'Arpenty (Essonne, France).

6. *SEM des Coccolithophoridées Fossiles*

du disque distal avec sa dépression centrale très accusée, le rapport des diamètres entre le disque distal et le disque proximal, la hauteur totale du coccolithe en relation avec son diamètre.

L'exemple suivant, *Cylindralithus biarcus* Bukry (1969) (Planche II, Fig. 1a, b, c, d) nous permet de tirer des renseignements sur la forme des éléments constitutifs que l'on peut suivre depuis la face proximale (Planche II, Fig. 1a) au niveau de laquelle ces éléments s'étalent en corolle jusqu'à leur extrémité distale acuminée (Planche II, Fig. 1d).

Cette possibilité de suivre les éléments constitutifs d'une face à l'autre du coccolithe revêt une grande importance pour notre compréhension de la construction générale des coccolithes. Le micropaléontologiste a très souvent tendance à traiter en parties indépendantes les différentes portions architecturales du coccolithe notamment au niveau de la couronne marginale ou de l'embase des corpuscules.

J'ai montré récemment (Noël, 1969) que les éléments constitutifs d'un coccolithe tel que *Gartnerago* Bukry (1969) (=*Laffittius*, Noël, 1969, Planche II, Fig. 3) pouvaient avoir une forme complexe que le microscope électronique à balayage permet de déceler. En effet, l'examen de la face proximale des coccolithes rapportés à ce genre décèle, au niveau de la couronne marginale, plusieurs couronnes concentriques emboîtées les unes dans les autres. En fait cette apparence est due à la forme particulière des éléments de cette couronne marginale que l'on peut suivre depuis l'aire centrale jusqu'au bord externe du coccolithe. Ces éléments juxtaposés avec leurs différents tronçons, déterminent des saillies et des dépressions que le microscope électronique à balayage met clairement en évidence.

Cette observation rejoint une observation publiée récemment (Watabé, 1967) sur *Coccolithus huxleyi*, observation modifiant profondément notre conception de la structure générale.

En outre, une étude de morphologie comparée (Noël, 1970b) m'a permis, grâce à des comparaisons élément par élément de mettre en évidence des affinités entre les deux genres *Gartnerago* et *Kamptnerius*. Affinités purement morphologiques certes mais qui témoignent vraisemblablement d'affinités biologiques et ne sont pas de simples convergences de forme.

CONCLUSION

Ainsi donc, ces quelques exemples choisis parmi les formes du Crétacé supérieur montrent l'intérêt du microscope électronique à balayage dans l'étude des nannofossiles. Par les précisions incontestables qu'il permet d'obtenir sur la morphologie de ces corpuscules, l'emploi de cet instrument nous conduit à une meilleure définition générique—et spécifique—des diverses formes rencontrées. Il devient possible d'étudier les différents caractères retenus comme critères en liaison les uns avec les autres, ce qui est d'une grande importance pour l'appréciation des variations individuelles au sein d'une même unité taxinomique.

En dehors de l'intérêt pratique pour les géologues d'avoir de bonnes définitions morphologiques des nannofossiles en vue de leur utilisation en tant que marqueurs stratigraphiques, le microscope électronique à balayage nous permet d'atteindre, d'un point de vue purement paléontologique, une connaissance plus approfondie de ces minuscules vestiges et de rechercher des liens phylétiques entre les différentes unités définies dans une optique strictement morphologique.

Enfin grâce à la possibilité que cet instrument nous offre de faire varier l'angle d'observation des corpuscules étudiés, il est relativement plus aisé qu'avec une réplique de carbone de paralléliser les images ainsi obtenues avec celles fournies par un microscope photonique. En effet lorsque l'on étudie en microscopie photonique des coccolithes dans un milieu de montage visqueux, il est possible d'examiner sous divers angles le même individu et d'avoir ainsi une meilleure idée de son architecture. Le transfert en microscopie électronique des renseignements fournis par le microscope photonique n'est pas négligeable, eu égard à l'importance de la littérature sur les coccolithes en microscopie photonique seulement.

REFERENCES

Bramlette, M. N. et Martini, E. (1964). The great change in calcareous nannoplankton fossils between the Maestrichtian and Danian. *Micropaleontology* **10(3)**, 291–322.

Bukry, D. (1969). Upper Cretaceous coccoliths from Texas and Europe. *The University of Kansas Paleont. Contrib.* **51** (2), 1–79.

Noël, D. (1965). *Sur les coccolithes du Jurassique européen et d'Afrique du Nord. Essai de classification des coccolithes fossiles.* 209 pp. Editions du CNRS, Paris.

Noël, D. (1969). Arkhangelskiella (Coccolithes crétacés) et formes affines du Bassin de Paris. *Rev. Micropal.* **11**, 191–204.

Noël, D. (1970a). *Coccolithes crétacés: la craie campanienne du Bassin de Paris.* 182 pp. Editions du CNRS, Paris.

Noël, D. (1970b). Eléments de morphologie comparée: *Gartnerago* et *Kamptnerius*. *Cahiers Micropal.* (à paraître).

Perch Nielsen, K. (1968). Der Feinbau und die Klassifikation der Coccolithen aus dem Maastrichtien von Dänemark. *Biol. Skr. Dan. Vid. Selsk.* **16/1**, 1–96.

Reinhardt, P. (1964). Einige Kalkflagellaten-Gattungen (Coccolithophoriden, Coccolithineen) aus dem Mesozoikum Deutschland. *Mber. Akad. Wiss. Berlin.* 749–759.

Reinhardt, P. (1966). Zur Taxinomie und Biostratigraphie des Fossilen Nannoplanktons aus dem Malm, der Kreide und dem Alterttertiär Mitteleuropas. *Freiberger Forschungshefte.* **196**, 1–109.

Watabe, N. (1967). Cristallographic Analysis of the Coccolith of *Coccolithus huxleyi*. *Calc. Tiss. Res.* **1**, 114–121.

7 | Scanning Electron Microscopy and Information Transfer in Systematic Micropaleontology

WILLIAM W. HAY

*Department of Geology, University of Illinois at Urbana, Illinois
and Division of Marine Geology and Geophysics,
Rosenstiel School of Marine and Atmospheric Sciences,
University of Miami, Florida*

Abstract: Scanning electron microscopy offers a unique opportunity to increase the rate of information transfer in micropaleontology. Because the biostratigraphic and ecologic interpretations derived from the study of microfossils are dependent on the accuracy of taxonomic identification, it has long been held that illustration of the specimens is invaluable. In some modern biostratigraphic work, however, it has become evident that ordinary illustrations are in themselves inadequate for identification, and the idea is current that with some special groups, such as the planktonic foraminifers, correct identification is possible only by direct comparison with the types. If this is true, it is an indication that the amount of information conveyed in the micrographs is too small to permit the necessary differentiation.

Experiments on the identification of microfossils with a conventional stereoscopic binocular microscope indicate that resolution better than 20 micra cannot be attained with incident illumination of calcium carbonate specimens immersed in air. For microfossils in the 1 mm size range (most foraminifers, ostracods, etc.) the information available using light is only about 2500 bits. Curiously, in publication that amount of information has been spread over 10,000 to 100,000 dots in screened half tone illustrations, giving the impression that much more information is present than is actually the case.

With slow scanning speeds and a 1000 line raster on a suitable cathode ray tube, the potential number of bits of information on a scanning electron micrograph is 1,000,000, or about 400 times that available to an observer using a stereoscopic binocular microscope. For most specimens, the actual increase in available information is less than the indicated potential, but is still more than one order of magnitude greater than that from light optical observations.

Because the quantity, quality, and characteristics of the information in scanning electron micrographs can be determined with precision, and can be deliberately modified to conform to standards by adjustment of line and frame rates and contrast, important information can be enhanced and irrelevant information suppressed. By reducing the quantity of information, but increasing its quality, considerable economies in illustration may be achieved. The use of new techniques such as deflection modulation and digital scanning to produce images having a different character but a high quality of information may open new avenues for systematic paleontology.

INTRODUCTION

The presently accepted methods of information transfer in systematic paleontology became stabilized during the first half of this century. The most widely approved technique consists of publication, preferably in a journal of broad distribution, of a series of taxonomic data including a synonymy, description, and figures of the fossils being discussed. The format in use today was devised at a time when there was little appreciation of the rapidity of organic evolution or of the sensitivity of organisms to minute environmental changes, and when the preparation of illustrations was a laborious and time-consuming task. The volume of paleontologic literature has increased at a very rapid rate, doubling about every ten years during this century. During the first half of this century, the increase in rate of publication was able more or less to keep pace with the increase in data, and many new journals were initiated during this period. During the past decade, however, the rate of production of data has increased so much that the traditional methods of publication as a means of information transfer have become woefully inadequate. Modern studies indicate that greatly refined biostratigraphic and paleoecologic interpretations are possible, but these require recognition of many more morphological types than have previously been used, and it is impossible for the traditional format for information transfer to handle the vast amounts of data now required for the progress of the science.

The Scanning Electron Microscope (SEM) offers unique opportunities for processing and quantifying information. It has become widely accepted as a means of illustration of microfossils, but the lessons in information theory which can be learned from the SEM extend far beyond its narrow application as a tool for illustration. The production of illustrations has changed from being the most difficult and time-consuming part of taxonomic work to the simplest, most rapid part, suggesting that it is now appropriate to consider whether the traditional techniques of information transfer in paleontology should not be altered to conform to the new realities (Sandberg and Hay, 1967, 1968; Hay and Sandberg, 1967).

In the following discussion, the reasons for vastly increasing the amount of illustration in the literature will be discussed first, followed by an inquiry into the relative information content in different kinds of illustration, an examination of the relative merits of illustration versus description, and finally some suggestions of alternative methods of information transfer.

THE USES OF PALEONTOLOGIC DATA

The cataloguing of fossil taxa is no longer regarded as an end in itself, but is important as a means of determining the ages of strata, studying the course of evolution, and interpreting the earth's climatic history.

The importance of fossils as a guide to the relative ages of rock strata has been recognized for a century and a half. Because the early development of the science of stratigraphy was intimately associated with the concepts of catastrophism and multiple creation, the field of biostratigraphy, which encompasses the determination of the relative ages of rocks by the use of fossils, has tended to express age equivalence or non-equivalence in terms of absolutes. The biostratigrapher commonly states that the rocks are of the same age, or that they are not of the same age; an expression of the degree of probability of age equivalence has rarely been offered. During the last century, correlation (which in the biostratigraphic sense refers to the establishment of age equivalence) was based largely on assemblages of fossils. More recently, it has become known that close resemblance between entire assemblages of fossils is more likely to reflect similarity of environmental conditions rather than absolute time equivalence. Biostratigraphic correlation by means of evolutionary series has replaced correlation based on similarity of assemblages but this technique approaches a limit of resolution of 1–3 million years in most groups of fossils. It should be understood that this is not a natural limit of significance in evolutionary theory, but is a practical limit set by the difficulties of distinguishing successive members of an evolutionary continuum. It commonly represents the point at which statistical studies must be initiated to differentiate between populations.

Hay and Čepek (1969) have suggested that a system of stratigraphy based on the probability of successive events could easily be established. Because this system will permit combination of datum levels from diverse evolutionary lineages, an order of magnitude increase over current biostratigraphic resolution should be attainable without increasing the number of taxa recognized. However, it seems possible that yet another order of magnitude increase in stratigraphic resolution (to the 10,000–30,000 year level) could be achieved by recognizing a larger number of morphological types. In any case the primary requirement of such a system is accuracy in the identification of morphological types or species. The resolution of the system is in large part a function of the latitude of the species or morphotype concept. In this case, as in much of science, resolution is proportional to the bandwidth with which the continuum is analysed. For consistency of results, it is of primary importance that the morphotype concepts be used by all workers in the same way, or at least that the concepts used in each laboratory be conveyed to others; this required a high volume of exchange of illustrative material, one to two orders of magnitude greater than is presently the case.

The contribution of paleontology to the study of evolution has been minimal because the stratigraphic order of the data is determined in large part by assuming certain evolutionary trends. Assumed evolution is the basis for much of our present biostratigraphy, but for objectivity interpretation of evolutionary lineages should be based on stratigraphic control. The considerable increase in the amount of published illustrative material required for the development of probabilistic sequential stratigraphic schemes would at last provide the necessary bank of data for objective investigation of evolutionary processes inherent in the fossil record.

Marszalek et al. (1969) have observed that many widely accepted concepts of paleoecology have been adapted from traditional biostratigraphy. The limits of the ranges of species are of the utmost importance in stratigraphy, and because the field of paleoecology was developed by biostratigraphers, the ranges of species were also assumed to be important in defining environmental boundaries. Ecologists have long regarded species dominance, not the limits of ranges, as fundamental to community classification. If paleontologic data are analysed according to the dominance of particular species or subspecies, a much more refined environmental classification is obtained. Examination of the data of Phleger (1960) on the distribution of foraminifers indicates that classification of foraminiferal communities based on dominant species would be at least twice as refined as that based on the limits of species ranges suggested by Phleger. The dominant species in adjacent communities usually differ markedly, rarely belonging to the same genus. However, over broad distances the same or closely related species occupy similar environments. As in the case of sequential stratigraphy, the resolution which can be achieved in differentiation of environments is a function of the bandwidth used to analyse the taxonomic spectrum. The more taxonomic units or morphotypes which can be reliably recognized, the more subtle the environmental differences which can be distinguished. For the refinement of paleoecology too, there is a need for greatly increased output of illustrations of the fossils found in different situations.

Emiliani (1969) has demonstrated the degree to which planktonic organisms can reflect minor environmental changes. He has shown that in the case of Pleistocene assemblages of planktonic foraminifers, the interaction of gradual evolutionary change and variation due to alteration of the environment during glacials and interglacials produces populations which are specific for a certain age and environmental condition. Verbal description of the degree of variation of size, shape, and texture characteristic for a particular population is virtually impossible, but illustration of suites of specimens of a particular species can adequately transfer the necessary information to an independent researcher. Emiliani observed that "Current paleontoligical research is handicapped by nomenclatorial formalism, the definition of species, and the designation of holotypes" (p. 265). A new way of distinguishing between closely related but different populations is required, and this again emphasizes the need for increased amounts of illustrative material in paleontologic information transfer.

INFORMATION IN ILLUSTRATIONS

Prior to the advent of the SEM, little attention was given to the amount of information contained in illustrations. With the SEM, however, the information in an illustration may be quantified, and it becomes necessary to decide what constitutes an adequate information level. The SEM has already been widely used in micropaleontology, particularly in the illustration of planktonic foraminifers. The most commonly heard criticism of scanning electron micrographs is that they contain too much information, and the observer is no longer sure of the identification of a species. The fact that too much information causes taxonomic confusion is an interesting paradox.

Micropaleontologists have traditionally viewed microfossils through a stereoscopic binocular microscope. The stereoscopic binocular microscope is designed to achieve maximum depth of field, and in accomplishing this end some resolution is sacrificed. Apparently, few micropaleontologists have realized just how poor the resolution of a stereoscopic binocular microscope really is. The theoretical limit of resolution of an optical microscope using visible light is about 0.25μ; this limit can actually be attained by viewing the object while it is embedded in a medium with a refractive index close to that of the object, using an oil immersion lens to achieve the greatest possible numerical aperture. The stereoscopic binocular microscope is made to view objects in a quite different manner. Micropaleontologists are accustomed to viewing the fossil in air, a medium with a refractive index far from that of the fossil. The resulting interference phenomena greatly reduce the resolution which can be attained, so that for calcite objects with a rough surface, observed by incident illumination and immersed in air, the resolution is only about 20μ, almost two orders of magnitude poorer than the theoretical limit. If the object is immersed in water, oil, or another medium with a refractive index closer to that of calcite, some improvement will be noted, but microfossils tend to become more transparent in these media, so that the diagnostic surface features used to identify the fossil tend to disappear.

Twenty microns are 1/50 mm, so that for a microfossil the size of most planktonic foraminifers, benthonic smaller foraminifers, of ostracods, which have a diameter of about 1 mm, it might be considered that the number of bits of information (i.e. discrete resolvable units of information) which the stereoscopic binocular light microscope can provide is only about 50×50 or 2500. The limit of resolution of the human eye is about 1/10 mm, but this is an absolute value and assumes optimum viewing conditions. Nevertheless, this suggests that an illustration of a microfossil enlarged to a diameter of 5 mm should suffice to convey all of the information available from observation with the stereoscopic binocular microscope. In practice, illustrations of microfossils have been published at two to four times this size, rendering them easier to view. The very fine screening used by some paleontological journals may represent the 2500 bits of information with 10,000 or even 100,000 printed dots, giving the impression that much more information is present than is actually the case. The problem of depth of field of light optics is apparent to any micropaleontologist who has tried to photograph microfossils using a light microscope. Hanna, in an interesting paper (1927) demonstrated that it is impossible to photograph some microfossils so that all the pertinent characters can be seen. He described experiments on the photomicrography of *Orbulina universa* d'Orbigny, a pelagic foraminifer with a spherical test up a 1 mm in diameter. Two sizes of pore are present in the test of *Orbulina universa*, and Hanna demonstrated that when resolution was adequate to show the smaller set of pores, depth of field was so shallow that only a very small part of the spherical test was in focus; conversely when most of the test was brought into focus by closing the aperture of the camera, the required resolution was lost.

With the SEM, the problem of depth of field is avoided, as the SEM shows an improvement of about 50 times over the stereoscopic light microscope for this factor. Further, by using a slow scanning speed and a 1000 line raster, the potential number of bits of information attainable is 1,000,000, or 400 times that which can be achieved with the stereoscopic binocular light microscope. It becomes easy to understand the criticism that "too much information is available". The amount of information available from the SEM is so much greater than that available previously that micropaleontologic speciation is often difficult. However, because micropaleontologists have utilized relatively little information in the recognition of fossils prior to the development of the SEM, it does not necessarily follow that it would be best to disregard the influx of information available from the SEM. It is clear that an "information gap" exists which must be bridged if older work is to be related to future work with the SEM, and this poses a very real problem. One way of bridging the gap is to systematically remove or reduce the amount of information in the scanning electron micrographs. The simplest way of reducing the amount of information is to remove it at regular intervals without regard for the relative value of different bits of information. This can be accomplished crudely by defocusing the beam so that the resolution begins to approximate that of the stereoscopic binocular microscope, or more elegantly by using a digital scanning system (such as that available on the AMR 900 SEM) with fewer scan points and a larger spot size. This technique may be of value to the individual investigator making the transition from the stereoscopic binocular microscope to the scanning electron microscope. Alternatively, the scanning electron micrographs may be covered by opaque sheets or patterned screens to reduce the amount of information available to the observer. Similar techniques have been employed by investigators working with calcareous nannoplankton fossils in order to make comparisons between light micrographs and transmission electron micrographs of carbon replicas of coccoliths.

Plate I, Fig. 1 is a drawing of *Globorotalia cultrata* (d'Orbigny) prepared by a highly competent observer and artist, Mrs. Barbara Lidz Miller, using a stereoscopic binocular microscope. The production of the drawing required many hours. Plate I, Fig. 2 is a scanning electron micrograph of another specimen of the same species, taken with a 20 second frame on Polaroid film so that the finished picture was available about one minute after the photographic recording began. The drawing is a more pleasant and aesthetically beautiful representation, but the scanning electron micrograph contains considerably more information. Plate II, Fig. 1 represents an enlargement of part of the same negative used to print Plate I, Fig. 2. Note that in the drawing it is impossible to render the pustules surrounding the umbilical region in any detail, but in the scanning electron micrograph they are seen to display crystal faces.

For information transfer to be optimal, it is important to utilize all pertinent bits of information; this means that the scale of the picture should be such that the structural details of the object are visible as well as its gross morphology and ornamentation. For planktonic foraminifers, the important structural features of the test are usually visible on the same scanning electron micrograph which displays the entire test. This means that a 10×10 cm illustration should suffice. Most previous illustrations of planktonic foraminifers have been printed at a scale so that the test covers an area of 4 cm^2 or less; the illustration size suggested here required an increase of 25 times in the space required to display a given suite of specimens. Investigation of the calcareous nannofossils is the only area of micropaleontology in which the use of the SEM does not imply a major increase in the size of illustrations and the space devoted to them in publications. For the past decade, the period during which the calcareous nannoplankton

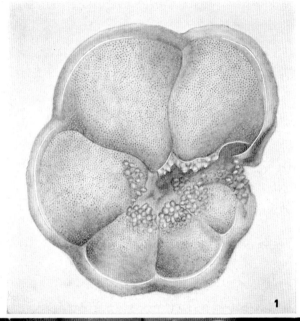

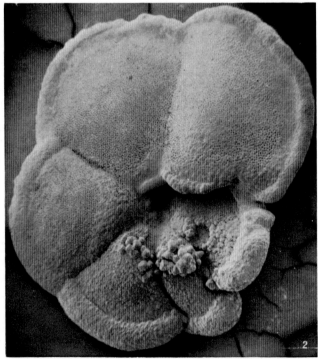

fossils have received intensive study, transmission electron micrographs of carbon replicas of the nannofossils have become accepted as the standard means of illustration, and the conversion to scanning electron micrographs required no increase in the amount of space required for illustration.

ILLUSTRATION VERSUS DESCRIPTION

During the last century, the preparation of illustrations was a major obstacle to the publication of monographic works. Consequently, illustrations tended to be few, and the verbal descriptions of fossils lengthy. With the development of photographic techniques, the problem of illustration became less formidable, but still required considerable effort. Today automatic exposure devices, Polaroid films, and rapid print processors are standard in most laboratories, so that the preparation of photographic illustrations is technically very easy and inexpensive. In the case of microfossils, the preparation of adequate illustrations by photographic techniques was difficult, and as noted above, in some cases impossible. The time required to produce suitable pictures, even with the best light microscopic equipment, is still very long compared with the SEM. Using Polaroid materials and a Leitz microscope modified after Triebel (1958), light micrographs suitable for publication could be obtained at a rate of about 1 satisfactory negative every 15 minutes, or about 32/day. This is a rate at least one order of magnitude faster than can be achieved by having an artist draw the fossils, but is about one order of magnitude slower than can be achieved with the SEM. Using Polaroid materials and a Stereoscan SEM, we have found it possible to routinely produce 280 negatives suitable for publication in an 8-hour working day. The SEM, then, makes it possible to produce one or two orders of magnitude more work with more than two orders of magnitude more information in the resulting pictures.

In spite of the ease with which illustrations can now be produced, and in spite of the reasons stated above why the amount of illustration in the published

---

Plate I

Fig. 1. *Globorotalia cultrata* (d'Orbigny), drawing by Mrs. B. L. Miller, $\times 80$, from Core P6804–003–03 (Lat. 17°55·5'N, Long. 65°06·1'W; Anegada Trough, Caribbean Sea), 4315 metres depth, Recent ?, Hypotype UI–H–5195. UI–H hypotypes are deposited in the collections of the Department of Geology, University of Illinois, Urbana, Illinois.

Fig. 2. *Globorotalia cultrata* (d'Orbigny), scanning electron micrograph, $\times 100$, from same locality as specimen in Fig. 1, Recent ?, Hypotype UI–H–5196.

literature should be drastically increased, the relative numbers of pages devoted to text and to illustration in most journals has not changed appreciably in the past 50 years. There seems to be an irresistible compulsion on the part of authors and editors to accompany each page of illustration with several pages of text describing the illustrations. The verbal description of illustrations is redundant, unnecessary, and should be discontinued. The only descriptions which should be allowed by editors are those which convey information not available from the illustrations, or available only by knowing relationships between different illustrations which are not obvious from the illustrations themselves. In many instances, this latter sort of relation can be expressed in the explanations of the illustrations, and may not be needed in the running text. Unfortunately, the requirement of the Codes of Zoological and of Botanical Nomenclature that new species be accompanied by an adequate description is often interpreted as meaning that elaborate descriptions of the illustrations are necessary. Modification of the rules is probably not needed, but liberalization of their interpretation would be welcome.

It can be easily demonstrated that the amount of information conveyed by a page of verbal description is infinitesimal compared with the information conveyed by a page of illustration. The reader can satisfy his curiosity on this point by attempting to sketch any fossil from its verbal description. Apparently it is tradition alone which required that published papers must have more pages of text than of illustration.

Editors cite the cost of a plate at 40% to 100% more than the cost of a printed page of text, justifying the current relative proportions of text to plates on this economic basis. The purpose of the journals is, however, to transfer information, not to produce a given number of printed pages each year. Illustrations are more efficient means of information transfer, and true economy lies in altering the ratio in favor of illustrations over text. The increased cost or reduced size of a journal resulting from its conversion from many pages of text and few plates to

---

Plate II

Fig. 1. Detail of Plate I, Fig. 2, made from same negative enlarged $\times 2 \cdot 5$ (magnification of specimen = $\times 250$). Shape of pustules on inner chambers of test determined by crystal faces. Shape of pores mostly indeterminate at this magnification as they have a diameter approximating the width of a single scan line. Note that the scan raster is visible over the entire print, indicating resolution of the CRT better than 1000 lines.

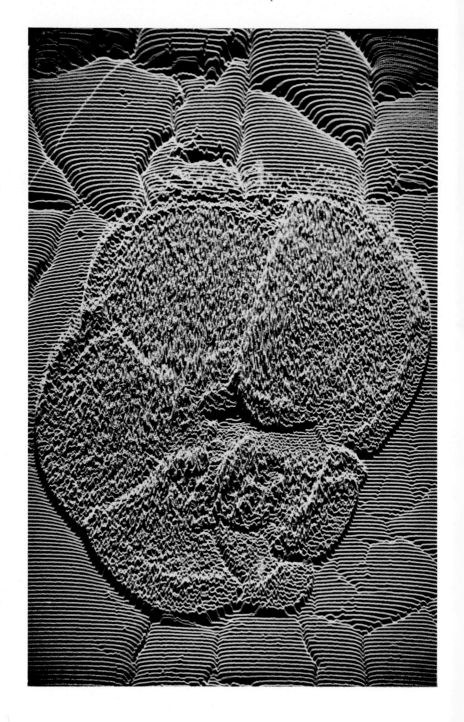

many plates and few pages of text would be more than offset by increased efficiency of information transfer.

The following specific suggestions for modification of the taxonomic sections of papers reflect present conditions:

1) Synonymies should be reduced; exhaustive lists serve little if any purpose. In many areas the form of the synonymies can be greatly abbreviated; for example, in the calcareous nannofossils only the species and authors names need be cited because detailed references to publications are available from the bibliographies and indices published by Loeblich and Tappan (1966, 1968, 1969, 1970).

2) Verbal description of illustrations should be eliminated; comments should be restricted to discussion of features not evident from the illustration. In the case of new species, the requirements of the codes can be fulfilled by citing those features which serve to differentiate the new species from previously published species.

3) Variability in a species should be described quantitatively, preferably in term of means and standard deviations presented in table form, or by means of illustration of typical and atypical members of the population.

4) The number of illustrations should be increased to adequately convey the basis for conclusions.

5) The method of reproduction of the illustrations should be appropriate to the quantity of information in the illustrations. The use of 1,000,000 printed dots to convey 2500 bits of information is wasteful and economically unfeasible.

6) The explanations of the figures should be complete and contain all necessary information (preferably in abbreviated form) regarding magnification, type of illustration, locality and stratigraphic level, repository number, etc. so that the illustration may serve as a basis for further study by other investigators. Consideration should also be given to the possibility of including all discussion of species in the explanation of the figures; this would eliminate extensive duplication and would shorten many systematic papers by half. Plate V illustrates how this might be done. This plate, illustrated with scanning electron micrographs is from a forthcoming publication by Wayne D. Bock, of the Rosenstiel School of Marine and Atmospheric Sciences of the University of Miami. The

---

Plate III

Fig. 1. Y-deflection modulated scanning electron micrograph of specimen illustrated in Plate I, Fig. 2. ×140. 200-line scanning raster.

paper deals with the distribution of living foraminifers in the Florida Bay, and includes a taxonomic section, describing the species present, a map showing the location of the stations studied, a table presenting the data on distribution, and a discussion of the results. The taxonomic section constitutes a substantial portion of the paper, but if the format suggested here were adopted and accepted by journals all of the text relating to taxonomy would be included in the plate explanations.

ALTERNATIVE METHODS OF INFORMATION TRANSFER

Publication of illustrations, diagrams, or words on paper constitutes the only method of information transfer widely accepted by paleontologists today. The recent innovation by P. C. Sylvester-Bradley of publishing scanning electron micrographs of ostracods with pertinent data but without interpretative commentary represents a significant departure from the traditional publication format used in journals and illustrates application of many of the suggestions mentioned above. Because SEM images are derived from electrical signals amenable to processing in a variety of ways, alternative methods of presentation of the information should be examined to determine whether more efficient, economical techniques of information transfer are possible.

As noted above, an SEM image may contain 1,000,000 or more bits of information. Not all of this information is useful however, as a certain noise level exists in all images. The most direct method of reducing the volume of data to be transferred without reducing the quality of the information is to minimize redundancy and noise. Reduction of the noise level in the signal generating components and circuitry of the SEM itself is a technical matter achieved by following manufacturers recommendations. Reduction of the noise level inherent in the specimen (imperfection due to poor preservation, preparation techniques,

---

Plate IV

Fig. 1. Composite scanning electron micrographs, showing same area illustrated in intensity modulated (above) and Y-deflection modulated modes. Half of the intensity modulated micrograph has been ruled by pen to indicate the areas from which the information presented in the Y-deflection micrograph is derived. By following the lines in the ruled area of the intensity modulated micrograph, and comparing the tonal variation with the fluctuations of the corresponding line in the Y-deflection modulated micrograph, an idea of the relative information content can be obtained. Variation in dark grey tone values appear very slight in the intensity modulated micrograph, and are more obvious in the Y-deflection modulated micrograph. ×250.

7. SEM and Information Transfer in Micropaleontology 137

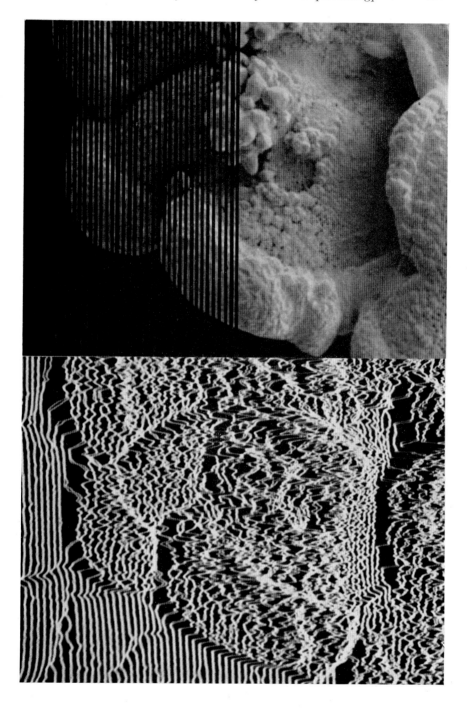

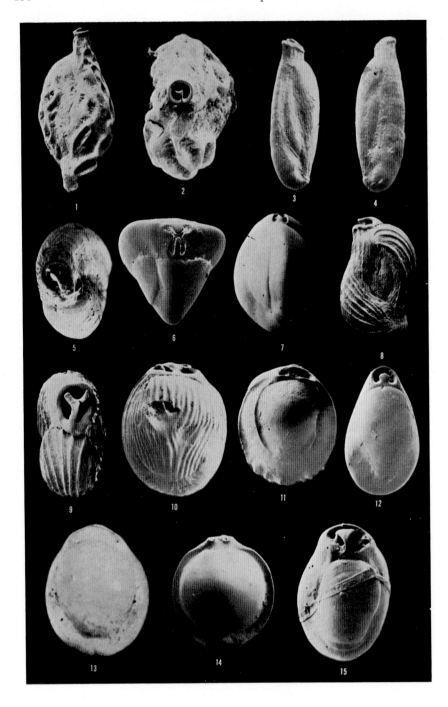

Plate V

Figs 1–2. *Quinqueloculina tricarinata* d'Orbigny, off Key West (24°38·7'N, 81°49·9'W), 6·2 m depth, USNM 643276.

Fig. 1. Side view, ×25.

Fig. 2. Apertural view ×70. Living specimens limited mainly to coarse sediment in euryhaline waters, but found occasionally on a vegetative substrate.

Figs 3–5. *Quinqueloculina* sp., Florida Bay (25°00·2'N, 80°37·4'W), 0·6 m depth, ML 154.

Figs 3 and 4. Side views, ×50.

Fig. 5. Apertural view, ×160. Living specimens limited to sediments in euryhaline waters.

Figs 6–7. *Cruciloculina triangularis* d'Orbigny, Gulf of Mexico (25°22'N, 82°01·6'W), 18·5 m depth, ML 55.

Fig. 6. Apertural view, ×125.

Fig. 7. Top view, ×85. Living specimens appear to be restricted to depths below 10 m, on fairly coarse sediment in stenohaline waters.

Figs 8–9. *Massilina protea* Parker, Florida Bay (25°00·2'N, 80°37·4'W), 0·6 m depth, USNM 643231.

Fig. 8. Side view, ×25.

Fig. 9. Apertural view, ×60. Living specimens occur in polyhaline lagoons and bays, usually on coarse sediment.

Fig. 10. *Pyrgo comata* Brady, Florida Straits (24°50'N, 80°03'W), 587 m depth, USNM 643252, top view, ×40. Living specimens mainly limited to depths below 25 m, on fairly coarse sediment in stenohaline waters.

Fig. 11. *Pyrgo denticulata* (Brady), Lower Florida Keys 24°43·9'N, 81°23·3'W), 0·3 m depth, USNM 643253, top view, ×45. Living specimens found on fine to coarse sediment in shallow, euryhaline waters.

Fig. 12. *Pyrgo elongata* (d'Orbigny), off Key Largo (25°00·8'N, 80°22·4'W), 6·5 m depth, USNM 643254, top view, ×75. Living specimens appear to be limited to fairly coarse sediment in stenohaline waters.

Fig. 13. *Pyrgo fornasinii* Chapman and Parr, Florida Straits (24°50'N, 80°03'W), 587 m depth, USNM 643255, top view, ×15. Living specimens usually limited to sediment substrate in fairly deep, stenohaline waters.

Fig. 14. *Pyrgo murrhina* (Schwager), off Key Largo (25°04·2'N, 80°23·7'W), 1·4 m depth, USNM 643256, top view, ×35. Living specimens occur on sediment in stenohaline waters.

Fig. 15. *Pyrgo subsphaerica* (d'Orbigny), Florida Bay (24°59'N, 80°23·3'W) 1·8 m depth, USNM 643257, top view, ×90. Living specimens found on both sediment and vegetative substrates in shallow euryhaline lagoons and bays. Band covering speimen is an artifact, probably a fragment of plant material.

All illustrations are scanning electron micrographs except Fig. 13, which is a light micrograph. All figured specimens are Recent, and are deposited in the collections of either the United States National Museum (USNM), Washington, D.C., or the Rosenstiel School of Marine and Atmospheric Sciences of the University of Miami (ML), Miami, Florida.

Plate V was made available by Dr. Wayne D. Bock, and is from a forthcoming publication, "Handbook of the Recent Benthonic Foraminifera of Florida Bay and Adjacent Waters". It is presented here with a modified explanation to demonstrate how the taxonomic text might be incorporated into the plate explanation. Synonymies and descriptions of the micrographs have been eliminated. Details of the distribution of the species illustrated are found on the maps and tables in Bock's publication.

etc.) required the skills of the trained observer coupled with facility in the operation of the instrument. The surface textures of objects observed in the SEM differ with the accelerating voltage. At 20kV the surface of a foraminiferal test may appear to be smooth and featureless, but at 5 kV it may appear to be finely textured; the micropaleontologist must decide whether the finely textured surface is worthy of note or whether it is merely an artifact of preservation or of the preparation technique. Determination of proper image qualities required experience with the SEM, and is one of the strongest arguments for the investigator also acting as operator of the instrument for the study of his specimens.

Use of the Y-deflection modulation to replace the usual intensity modulation permits examination of the structure of the image. It is surprising how often pictures made with Y-deflection modulation can be recognized and even identified correctly to subspecies. The number of scans in such pictures is relatively few, usually no more than 100, and some distortion of the shape of the object results. Nevertheless, enough information is retained so that the objects are readily recognized. In this case Plate III, Fig. 1 is a Y-deflection modulated scanning electron micrograph of the same specimen seen in Plate I, Fig. 2 and Plate II, Fig. 1. In this case, it is the redundant information which has been removed although in the process the other information has been somewhat distorted. Close examination of Y-deflection modulation pictures shows that each scan line has long segments which are essentially flat and shorter sections are inclined, but straight. Plate IV, Fig. 1 is a composite, showing an enlarged part of the scan raster in the umbilical area of the foraminifer illustrated in Plate I, Fig. 2 and Plate II, Fig. 1, and the Y-deflection modulation of the same region. Where the signal is flat in the Y-deflection modulated image, the signal is constant; in the regions where the segments of the Y-deflection modulation image are inclined but straight, the signal is changing at a constant rate. D. Raup (personal communication) has suggested that the path of the line could be determined by noting only the peaks and troughs. Figure 1 is derived from Plate III, Fig. 1 by marking each peak and valley by a dot. Figure 1 was prepared by hand by Mrs. B. L. Miller, and required considerable time, but data reduction of this sort could be accomplished electronically. This has not yet been done, but if carried out electronically on a normal intensity modulated raster, the result should resemble a point drawing.

Data reduction constitutes one phase of improving the economy and efficiency of information transfer; the method of transfer constitutes the other half of the problem. Printing on paper is likely to remain the most common method for at least the next decade. Signals from the SEM are just as readily stored on magnetic tape as on photographic film, however, and in some commercially available SEMs provision has been made for digital scanning which could be coupled directly to a computer for storage or processing.

With the advent of devices for transfering information stored on tape or discs in computers through telephone lines, it is not unreasonable to forecast that a time will come when there will be central memory banks for taxonomic illustration, and the individual researcher could obtain his data efficiently from this

Fig. 1. Point drawing made by marking the peaks and troughs indicated on Plate III, Fig. 1. This drawing was prepared by hand, but similar point drawings could be made electronically from intensity modulated images.
×140.

source, thus avoiding the painful search of the literature. Although such utopian ideas have no immediate prospect of fulfilment, more modest beginnings are appearing. The University of Miami is starting a data bank for the calcareous nannoplankton in an effort to stabilize identification and to arrive at a system of probabilistic sequential stratigraphy. Sylvester-Bradley's experiment in communicating among ostracod workers represents another new approach. Taxonomy now enters a new phase; it has become established as the ultimate basis for stratigraphic and ecologic studies, and the demands placed on it will be much greater than ever before. Technological and theoretical advances permit new methods to be used so that radical solutions to old problems become feasible.

ACKNOWLEDGEMENTS

Figure 1 and Plate I, Fig. 1 were prepared by Mrs. Barbara Lidz Miller, to whom the author is greatly indebted. Plate V was kindly provided by Dr. Wayne D. Bock, whose permission to allow its use as an example of the possibility of combining taxonomic observations and plate explanations is very much appreciated. The scanning electron micrographs were taken on a Stereoscan Mark IIA instrument acquired with funds from the National Science Foundation (GA–1239), Public Health Service (PH–FR–07030) and the Universtity of Illinois Research Board. The instrument is housed in the Central Electron Microscope Laboratory of the University, and to the director of that facility, Professor B. Vincent Hall, I express thanks for the excellent maintenance of the SEM. Mr. Lee Dreyer and Mrs. Olive Stayton aided in technical matters.

This work was supported by National Science Foundation Grant GA–15261.

REFERENCES

Bock, W. D. (In press). "Handbook of the Recent Benthonic Foraminifera of Florida Bay and Adjacent Waters".

Emiliani, C. E. (1969). A new paleontology. *Micropaleontology* **15**, 2658300.

Hanna, G. D. (1927). The photography of small objects. *Trans. Am. Microsc. Soc.* **46**, 15–25.

Hay, W. W. and P. Čepek (1969). Nannofossils, probability, and biostratigraphic resolution. *Bull. Am. Ass. Petrol. Geol.* **53**, 721.

Hay, W. W. and P. A. Sandberg (1967). The scanning electron microscope, a major breakthrough for micropaleontology. *Micropaleontology* **13**, 407–418.

Loeblich, A. R. Jr. and H. Tappan (1966). Annotated index and bibliography of the calcareous nannoplankton. *Phycologia* **5**, 81–216.

Loeblich, A. R. Jr. and H. Tappan (1968). Annotated index and bibliography of the calcareous nannoplankton. *J. Paleontol.* **42**, 584–598.

Loeblich, A. R. Jr. and H. Tappan (1969). Annotated index and bibliography of the calcareous nannoplankton III. *J. Paleontol.* **43**, 568–588.

Loeblich, A. R. Jr. and H. Tappan (1970). Annotated index and bibliography of the calcareous nannoplankton IV. *J. Paleontol.* **44**, 558–574.

MARSZALEK, D. S., R. C. WRIGHT and W. W. HAY (1969). Function of the test in foraminifera. *Trans. Gulf Coast Ass. Geol. Soc.* **19**, 341–352.

PHLEGER, F. B. (1960). "Ecology and Distribution of Recent Foraminifera". 297 pp, Johns Hopkins Press, Baltimore.

SANDBERG, P. A. and W. W. HAY (1967). Study of microfossils by means of the scanning electron microscope. *J. Paleontol.* **41**, 999–1001.

SANDBERG, P. A. and W. W. HAY (1968). Application of the scanning electron microscope in paleontology and geology. *Proc. Symp. Scan. Elec. Microsc., Instr. and Appl.* pp. 29–38, IIT Res. Inst., Chicago.

TRIEBEL, E. (1958). Die Photographie im Dienste der Mikropalaeontologie. In "Handbuch der Mikroskopie in der Technik", (Freund, H., ed.) 2: Pt. **3**, pp. 83–144. Umschau Verlag, Frankfurt am Main.

8 | Interprétation Botanique des Pollens Inaperturés du Mésozoïque Saharien. Essai de Classification d'après l'Observation en Microscopie Électronique à Balayage

Y. REYRE
Laboratoire de Géologie du Muséum d'Histoire Naturelle, Paris

Abstract: The taxonomic value of the sculpturing of the exine of pollen grains of present-day Gymnosperms has been demonstrated by scanning electron microscopy. Study of such sculpturing on inaperturate dispersed pollen from the Mesozoic of the Sahara and Israel allows some of them to be attributed to present-day families (Taxaceae, Cupressaceae, Araucariaceae) while others, showing particular kinds of sculpturing, cannot be so placed and must represent unknown fossil taxa. These observations also indicate that the morphographic nomenclature of inaperturate pollen grains based on the features seen by light microscopy (shape, structure, size, surface appearance) corresponds to an artificial classification. The nomenclature proposed here is a combination of certain traditionally accepted form genera and new organ genera aimed at indicating, whenever possible, the probable botanical source taxon. In view of the inadequacy of the diagnoses based on light microscopy, which in the absence of a precise description of the exine sculpturing, *cannot adequately characterize a species*, it is necessary to provide new specific diagnoses for forms which correspond by light microscopy to validly published, but inadequately described species, as well as describing new forms.

Résumé: L'observation au microscope électronique à balayage de pollens de Gymnospermes actuelles a permis de reconnaître la valeur taxonomique de la sculpture externe de l'exine (Y. Reyre, 1968). L'analyse de ce caractère sur les pollens inaperturés dispersés du Mésozoïque du Sahara et d'Israël permet d'attribuer certains d'entre eux à des familles actuelles (Taxacées, Cupressacées, Araucariacées) tandis que d'autres, présentant des types de sculpture particuliers, ne peuvent que révéler l'existence de taxons fossiles également particuliers et inconnus. Ces observations mettent en outre en évidence que la nomenclature morphographique des pollens inaperturés telle qu'elle résulte des seuls caractères observables en microscopie photonique (forme, structure, taille, apparence imprécise ou déformée de la sculpture) correspond à une classification artificielle. Toutefois une classification naturelle (botanique) ne saurait être fondée sur le seul caractère sculptural; en effet certains types de sculpture se retrouvent sur des pollens de forme et de structure différentes. La nomenclature proposée dans la présente communication est une combinaison de certains genres de forme (form genus) classiquement admis et de genres d'organes (organ genus) nouveaux destinés à désigner, lorsque c'est possible, le taxon botanique producteur probable. Toutefois devant l'insuffisance des diagnoses établies en microscopie photonique, diagnoses qui, en raison de l'absence d'une description précise de la sculpture, *ne sont pas à même de caractériser une espèce*, il est nécessaire de procéder à de nouvelles diagnoses spécifiques soit à partir de formes correspondant en microscopie photonique a des espèces valablement publiées mais insuffisamment décrites, soit à partir de formes nouvelles.

INTRODUCTION

Les sédiments du Jurassique et du Crétacé inférieur du Sahara et d'Israël contiennent de grandes quantités de pollens inaperturés souvent difficiles à classer par l'observation en microscopie photonique. La systématique établie sur les seuls caractères observables par ce moyen est en effet très confuse en raison du fait que les auteurs ont tenu compte pour établir les diagnoses génériques de hiérarchies des caractères différentes selon chacun. Les uns considèrent comme le caractère primordial la dimension, d'autres la forme, d'autres l'épaisseur de l'exine, d'autres la sculpture apparente en microscopie photonique, d'autres enfin l'éventuelle aperture (monoporés). De ce fait les diagnoses génériques se superposent au moins partiellement, rendant très délicate l'attribution d'une forme à tel ou tel genre morphographique.

Afin de chercher à clarifier l'importance botanique des divers caractères chez ces pollens inaperturés appartenant pour la majorité à des Gymnospermes fossiles, j'ai observé au microscope électronique à balayage d'une part toute une série de pollens appartenant à toutes les familles actuelles de Gymnospermes, d'autre part des espèces dispersées dans les sédiments mésozoïques sahariens et attribuables en microscopie photonique aux genres morphographiques les plus courants et quantitativement les plus importants. Cette comparaison a conduit à envisager une classification moins artificielle des pollens dispersés.

Je rappellerai donc d'abord la nomenclature en microscopie photonique telle qu'elle paraît la plus généralement admise dans la bibliographie actuelle. Puis je rappellerai les principales constatations résultant de l'observation en microscopie à balayage des pollens de Gymnospermes actuelles. Ensuite je montrerai les principaux types de sculpture observés sur les pollens dispersés du Mésozoïque et leurs relations avec les caractères structuraux. Ceci m'amènera à exposer, d'une manière simplifiée, les éléments d'une nomenclature nouvelle "paranaturelle" que j'ai employée dans un mémoire intitulé "Palynologie du Mésozoique saharien" (thèse présentée à la Faculté des Sciences de Paris, 1970, à paraître) pour classer les pollens d'après la sculpture de l'exine.

Les termes employés pour la description des sculptures de l'exine ont tous été définis en langue française dans une précédente publication (Reyre, 1968c).

LA CLASSIFICATION ACTUELLE DES POLLENS INAPERTURÉS EN MICROSCOPIE PHOTONIQUE

Je ne m'étendrai pas ici sur l'étude historique et critique des définitions génériques ni sur l'acception de ces genres telle qu'elle apparaît à l'usage dans la bibliographie; je me contenterai d'en donner un résumé sous forme de tableau en spécifiant toutefois les auteurs originaux ou bien les auteurs d'amendement ainsi que dans certains cas l'acception générale du genre (*sensu lato*) lorsqu'elle est différente de la diagnose précisément définie par un auteur.

I. Pollens de grande taille (plus de 45 μ)
  – exine lisse: peu plissés — *Inaperturopollenites sensu lato*
        plissés — *Laricoïdites* Potonié, Thomson & Thiergart, 1950
        très plissés — *Psophosphaera* Naumova, 1950 Potonié, 1956
  – exine finement granuleuse — *Inaperturopollenites sensu lato*
  – exine fortement granuleuse — *Araucariacites* Cookson, 1947

II. Pollens de petite taille (jusqu'à 40–45 μ)
  – aletes (sans aperture)
        plissés à exine lisse — *Inaperturopollenites* (Pflug, 1952 ex Thomson & Pflug, 1953) Potonié,
        plissés à exine granuleuse — indésignables
        non plissés ou lenticulaires granuleux ou lisses — *Sphaeripollenites* Couper, 1958 Jansonius, 1962
  – monoporés — *Sphaeripollenites* Couper, 1958 Burger, 1966

III. Pollens à plis ou structures concentriques mais sans vésicules ni corps central net — *Inaperturopollenites sensu lato*

IV. Pollens de forme ellipsoïdale et munis de rayures fines ou plis parallèles à l'axe polaire du grain — *Leioaletes* Staplin, 1960
        ou *Podozamites* Bolchovitina

Tous ces genres palynologiques sont pour la plupart artificiels sous l'angle de l'appartenance botanique des espèces qu'ils désignent. Ceci est déjà révélé dans certains cas par des considérations structurales: Gamero (1965) indique que le macrofossile *Apterocladus* a fourni un pollen *in situ* appartenant au genre *Tsugaepollenites*; de même Townrow (1967) précise que le macrofossile *Masculostrobus warrenii* a également un pollen *in situ* de ce type. Or ces deux macrofossiles seraient des Podocarpacées. Au contraire l'espèce secondaire et tertiaire *Tsugaepollenites mesozoïcus* est très proche structuralement des pollens de *Tsuga* actuels. Ce caractère artificiel de la nomenclature actuelle est nettement perçu par l'observation de la sculpture en microscopie électronique. J'ai montré (Reyre, 1968b) comment trois espèces du genre de forme *Sphaeripollinites* Couper, 1958 pouvaient être d'après la sculpture attribuées à trois taxons botaniques différents, dont deux étaient assimilables à des familles botaniques actuelles. Avant d'approfondir cette question, je voudrais rappeler quelques constatations faites sur les pollens de Gymnospermes actuelles.

OBSERVATION DES POLLENS DE GYMNOSPERMES ACTUELLES
AU MICROSCOPE A BALAYAGE

De précédentes publications (Reyre, 1968a, c) ont traité de ce sujet et je rappelle brièvement les principaux résultats. Quatre sortes de dispositions de la sculpture de l'exine sont distinguables (Reyre, 1968c, Tableau I, p. 202): sculpture simple, composée, double et crépie. A l'intérieur de chacune de ces dispositions, on peut discerner plusieurs types élémentaires de sculpture selon la nature des éléments constituants. Ceux-ci permettent en général de caractériser palynologiquement un groupe de genres de la même famille, voire d'espèces du même genre; de plus lorsqu'on fait appel à des mesures dimensionnelles et quantitatives des éléments de sculpture, on peut en général caractériser une espèce précise. Nous devons donc être guidés par ces considérations dans la définition des espèces dispersées. Enfin il apparait, dans l'état actuel des connaissances, que dans chaque famille de Gymnospermes les espèces présentent un même type de sculpture et que certaines familles ont un type de sculpture particulier.

Il faut d'autre part noter un fait observable chez les Araucariacées actuelles. *Araucaria bidwelli* Hook. par exemple a un pollen dont l'exine fine se plissant facilement (Reyre, 1968c, Planche II, Fig. 4) alors que celui de l'*Araucaria*

Planche I

Vues générales des formes ou espèces (grandissement ×1000).
Fig. 1. *Taxacites sahariensis*.*
Fig. 2. cf. *Sphaeripollenites scabratus* Couper.
Fig. 3. *Cupressacites oxycedroïdes*.*
Fig. 4. *Cupressacites* sp. 2.*
Fig. 5. cf. *Inaperturopollenites giganteus* Goczan.
Fig. 6. *Inaperturopollenites* cf. *Araucariacites australis* Cookson.
Fig. 7. *Araucariapollenites* sp. 2.*
Fig. 8. *Inaperturopollenites* cf. *Araucariapollenites*.
Fig. 9. cf. *Zonalapollenites segmentatus* Balme.

8. *Interprétation Botanique des Pollens du Mésozoïque Saharien* 149

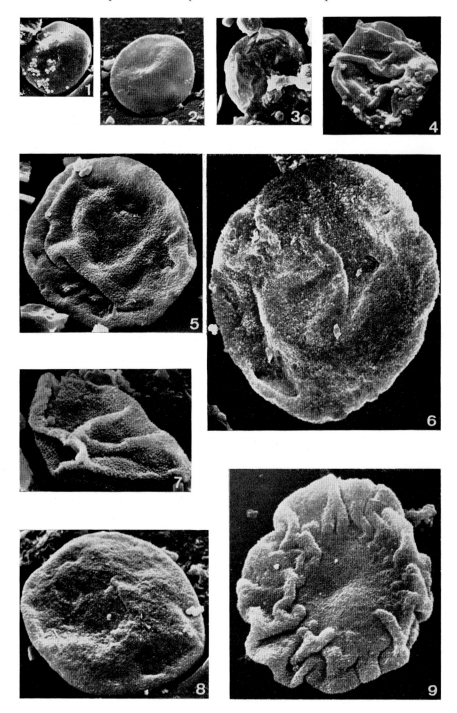

brasiliensis Rich. ne se plisse pas ou peu en raison, semble-t-il, d'une exine fortement structurée (Reyre, 1968c, Planche II, Fig. 5); le premier a d'ailleurs l'allure d'un sac originellement sphérique tandis que le second est lenticulaire. Nous verrons ci-dessous que de phénomène est également observé chez les pollens fossiles mais dans le cas présent il revêt une importance particulière parce qu'il s'agit d'espèces appartenant avec certitude au même genre botanique. Ainsi, *tandis que la forme générale et l'épaisseur de l'exine varient dans un même genre, la sculpture de l'exine reste du même type* et ce caractère apparait alors d'une importance primordiale pour l'identification des pollens fossiles dispersés et leur classement dans une nomenclature "para-botanique".

PRINCIPE D'UNE SYSTÉMATIQUE "PARA-NATURELLE"

L'observation des pollens inaperturés du Mésozoïque saharien révèle une assez grande diversité de sculptures, mais celles-ci ne sont pas liées à un type particulier de structure générale du grain. Ce fait a déjà été observé sur les pollens de Gymnospermes actuelles. Ainsi l'apparence générale du grain de pollen inaperturé, directement liée à sa forme, globulaire ou lenticulaire, et à la rigidité et l'épaisseur de son exine (pollens plissés ou non plissés) ne semble pas pouvoir être utilisée pour établir une systématique para-naturelle, c'est-à-dire conçue de manière à regrouper les espèces palynologiques selon leur appartenance probable à des taxons paléobotaniques voisins. Au contraire si l'on en juge par les résultats provenant des pollens inaperturés actuels la sculpture externe de l'exine serait un caractère primordial pour l'établissement d'une systématique para-naturelle. Toutefois ce caractère ne peut être considéré exclusivement: certains groupes de pollens dont la structure générale commune est si particulière qu'elle implique une provenance probable également commune montrent pourtant une évidente

---

Planche II

Détails montrant la sculpture de l'exine des formes ou espèces (grandissement $\times$ 10,000).
Fig. 1. *Taxacites sahariensis.**
Fig. 2. cf. *Sphaeripollenites scabratus* Couper.
Fig. 3. *Cupressacites oxycedroïdes.**
Fig. 4. *Cupressacites* sp. 2.*
Fig. 5. cf. *Inaperturopollenites giganteus* Goczan.
Fig. 6. *Inaperturopollenites* cf. *Araucariacites australis* Cookson.
Fig. 7. *Araucariapollenites* sp. 2.*
Fig. 8. *Inaperturopollenites* cf. *Araucariapollenites.*
Fig. 9. cf. *Zonalapollenites segmentatus* Balme.

8. Interprétation Botanique des Pollens du Mésozoïque Saharien 151

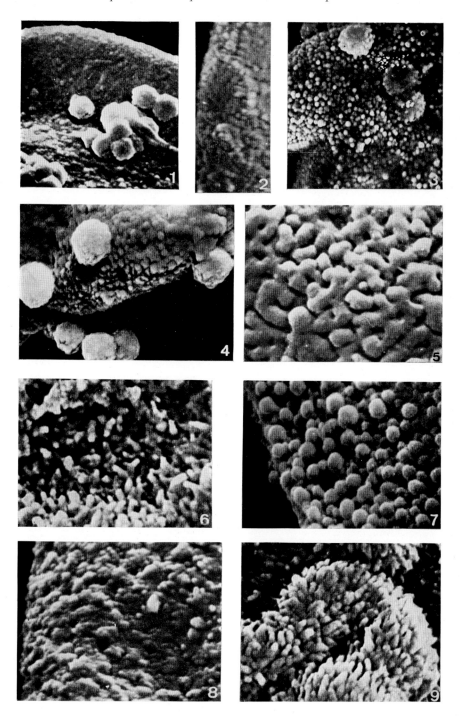

variabilité de la sculpture (*Classopollis, Zonalapollenites*). De ce fait la systématique para-naturelle des pollens inaperturés fossiles ne peut être qu'un compromis entre d'une part les considérations structurales observées en microscopie photonique et les caractéristiques sculpturales observées en microscopie à balayage. Le principe adopté est le suivant: lorsqu'un pollen a une sculpture évoquant celle observée dans une des familles actuelles il est rangé dans un genre morphologique nouveau évoquant également la famille considérée. Lorsque sa sculpture n'est caractéristique d'aucune famille actuelle il est attribué à un morphogenre classiquement admis.

Remarque: l'interprétation des sculptures telles qu'elles se présentent sur les grains fossiles extraits des sédiments par des traitements mécaniques et chimiques violents est parfois délicate. Certaines sculptures sont partiellement ou totalement détruites. Par exemple telle sculpture simple à boule peut se réduire à quelques boules éparses sur une surface lisse ou à peine rugueuse. Pour mesurer la densité des éléments de sculpture il faut choisir les plages où ils sont les plus contigus ou rapprochés. L'interprétation de la sculpture originelle reste parfois impossible.

LES GENRES

Sur la Planche I sont figurés 9 grains de pollen ($\times 1000$) et sur la Planche II les sculptures correspondantes ($\times 10,000$). Certains de ces grains correspondent à des genres ou des espèces dont les diagnoses originales sont établies dans le mémoire sur la "Palynologie du Mésozoïque saharien" (thèse soutenue à la Faculté des Sciences de Paris, 1970, *a paraître*); d'autres ne peuvent qu'être conférés à des espèces classiques. Les quelques exemples retenus dans la présente note ont été choisis pour expliquer et illustrer la nomenclature employée dans ce mémoire.

Dans le genre *Taxacites* sont rangés les pollens dont l'exine a une sculpture simple, verruqueuse, mamelonnée ou à boules, les éléments de sculpture, de densité généralement forte, reposant sur une surface basale lisse ou rugueuse. La taille de ces pollens est souvent petite et la forme lenticulaire ou sphérique, taille et formes fréquentes chez les pollens de Taxacées actuelles.

Un specimen de l'espèce-type de ce genre, *Taxacites sahariensis* Reyre est figuré Planche I, Fig. 1 et Planche II, Fig. 1. Il s'agit d'un pollen fossile dont la sculpture est voisine de celle des Taxacées actuelles (Reyre, 1968c) d'où le nom générique *Taxacites*.

Le genre *Cupressacites* regroupe les pollens dont la sculpture de l'exine est double, hétéromorphe ou isomorphe à glomérules reposant sur une couche basale memelonnée, verruqueuse ou parsemée de boules. La taille de ces pollens est souvent petite et la forme sphérique, ils peuvent avoir un pore, une marque trilète ou une scissure longitudinale ; tous ces caractères se rencontrent également chez les Cupressacées.

Un specimen de l'espèce-type *Cupressacites oxycedroides* Reyre (Callovien-Kimmeridgien) est figuré Planche I, Fig. 2 et Planche II, Fig. 2 ; un autre pollen de ce genre est figuré Planche I, Fig. 3 et Planche II, Fig. 3. Il correspond à la forme désignée *Cupressacites* sp. 2.

Il s'agit de pollens dont la sculpture est voisine de celle des pollens de Cupressacées actuelles.

Le genre *Araucariapollenites* a été créé pour regrouper les pollens inaperturés dont l'exine a une sculpture crêpie, cette apparence étant due soit à des éléments autonomes superposés soit à des éléments rattachés à la couche basale (sorte de clavules à tête verruqueuse ou en forme de boule).

Ce genre est ici illustré Planche I, Fig. 7 et Planche II, Fig. 7 par un grain appartenant à la forme désignée *Araucariapollenites* sp. 2 (Lias-Dogger inférieur), forme dont la sculpture ressemble beaucoup à celle des pollens d'Araucariacées en général et d'*Araucaria brasiliensis* en particulier. Mais cette forme n'a pas été trouvée en quantité suffisante dans un même échantillon pour autoriser la création d'une espèce.

La sculpture du pollen figuré Planche I, Fig. 8 et Planche II, Fig. 3 représenterait un cas limite pour l'attribution au genre *Araucariapollenites* : il s'agit d'un cas intermédiaire entre une sculpture crépie et une sculpture verruqueuse (pollen du Bathonien).

Un certain nombre de pollens ont une sculpture qui, dans l'état actuel des observations, n'a pas été encore reconnue chez les Gymnospermes actuelles. Dans ce cas les pollens sont attribués aux genres de forme tels qu'ils sont définis en microscopie photonique à l'exclusion du genre *Araucariacites* Cookson afin d'éviter les confusions.

Planche I, Fig. 2 et Planche II, Fig. 2 est représenté un specimen d'une forme très répandue dans le Jurassique du Sahara. Sa sculpture n'est pas caractéristique d'un des trois genres créés (dans ce cas seuls les caractères de morphologie générale sont considérés) ; en raison de sa petite taille, de sa forme lenticulaire et de l'apparence scabre de son exine en microscopie photonique il a été désigné cf. *Sphaeripollenites scabratus* Couper.

Planche I, Fig. 5 et Planche II, Fig. 5 est représenté un pollen de l'Albien conféré à l'espèce *Inaperturopollenites giganteus* Goczan : sa sculpture très particulière, de type cérébroide, n'est pas connue chez les pollens actuels et aucune interprétation botanique ne peut être proposée ; l'espèce est maintenue dans le genre *Inaperturopollenites*.

Planche I, Fig. 6 et Planche II, Fig. 6 est représenté un pollen du Jurassique supérieur attribuable d'après l'apparence en microscopie photonique à l'espèce *Araucariacites australis* Cookson; la sculpture échinulée est encore inconnue chez les pollens actuels d'Araucariacées et ce pollen a été désigné *Inaperturopollenites* (sensu lato) cf. *A. australis* Cookson.

Enfin Planche I, Fig. 9 et Planche II, Fig. 9 est représenté un type de pollen inaperturé que les considérations structurales excluent *a priori* des trois genres créés : la forme et la structure très particulières (vésicule entourant un corps central) sont dans ce cas considérés comme des caractères génériques plus importants que la sculpture de l'exine : ces pollens ont été classés dans le genre de forme *Zonalapollenites* Pflug ; il s'agit ici de l'espèce *Z. segmentatus* Balme.

CONCLUSION

La nouvelle nomenclature résumée ici a été conçue de manière à tirer partie d'un critère taxonomique nouveau, celui de la sculpture de l'exine. Elle ne rend caduque aucun des principaux genres créés en microscopie photonique à l'exception peut-être du genre *Araucariacites* Cookson qu'il est préférable de ne pas utiliser, car les pollens paraissant grenus en microscopie photonique ne sont pas tous attribuables à des Araucariacées fossiles. Enfin cette nomenclature elle-même devrait être modifiée dans le sens de la précision au fur et à mesure que progresseront les connaissances sur la signification botanique de la morphologie, de la structure et de la sculpture des pollens.

BIBLIOGRAPHIE

Balme, B. E. (1957). Spores and Pollen grains from the Mesozoic of Western Australia. *Commonwealth Sci. Ind. Res. Org., Coal Res. Sect.*, 48 pp., 11 pls.

Couper, R. A. (1958). British mesozoïc microspores and pollen grains. *Palaeontogr.* Abt. B, **103**, Liefg. 4–6, 175–179.

Reyre, Y. (1968a). Valeur taxinomique de la sculpture de l'exine des pollens de Gymnospermes et de Chlamydospermes. *C.r. Acad. Sci., Paris D.* **267**, 160–162.

Reyre, Y. (1968b). Valeur taxinomique de la sculpture de l'exine des pollens fossiles attribués aux Gymnospermes ou aux Chlamydospermes. *C.r. Acad. Sci., Paris, D.* **267**, 488–490.

Reyre, Y. (1968c). La sculpture de l'exine des pollens des Gymnospermes et des Chlamydospermes et son utilisation dans l'identification des pollens fossiles. *Pollen et Spores* **10**, 197–220.

9 | Generic Limits in the Biddulphiaceae As Indicated by the Scanning Electron Microscope

R. ROSS and PATRICIA A. SIMS

Department of Botany, British Museum (Natural History), London, England

Abstract: A brief review of the history of the generic classification of the Biddulphiaceae is given. The classification that is currently accepted by most workers recognizes a number of small genera and two large ones, *Biddulphia*, with bipolar valves, and *Triceratium* with multiangular valves. The structure of ten species as seen under the scanning electron microscope is described, three currently placed in *Biddulphia*, including its type and the type of *Zygoceros*, six currently placed in *Triceratium*, including its type and the types of *Amphitetras* and *Amphipentas*, and the type of *Cerataulus*. The results show that the division between bipolar and multiangular species separates species closely similar in other respects and unites species that are very different. Tentative suggestions are made about more correct generic limits that require examination of more species for their confirmation. The valves of some species are porose, others alveolate; these terms are defined. In some species the structure of the processes grades into that of the main part of the valve; in others the processes have perforate plates similar in structure to the ocelli of *Auliscus* and related genera.

INTRODUCTION

Before the introduction of the scanning electron microscope the senior author (Ross, 1963) suggested that the taxonomy above the specific level in the Biddulphiaceae was patently unsatisfactory. Accordingly, when the instrument did become available, we turned our attention to a survey of the fine structure of the family, and particularly of the species currently placed in the two large genera, *Biddulphia* Gray and *Triceratium* Ehrenb., that together include the greater part of the family as a whole. Our coverage so far is not complete enough for our conclusions to be more than tentative, but the observations we have already made do show that considerable revisions in generic limits will be required.

A brief historical review of the classifications proposed for the group will show that there have throughout been differences of opinion about its proper systematic arrangement. The earliest that need concern us is that of Roper (1859). He pointed out that Ehrenberg recognized four different genera of bipolar forms:

> *Biddulphia* Gray, united in chains, without spines;
> *Denticella* Ehrenb., a superfluous name for *Odontella* Ag., united in chains, with spines;
> *Zygoceros* Ehrenb., cells solitary, without spines;
> *Cerataulus* Ehrenb., cells solitary, with spines;

but maintained that these distinctions were unimportant and united all these genera in one, *Biddulphia*. He discussed the relation of this genus to the multi-angular forms but continued to accept a classification of these into genera based on the number of angles, maintaining *Triceratium*, with three angles, *Amphitetras* Ehrenb., with four angles, and *Amphipentas* Ehrenb., with five angles.

Ralfs (in Pritchard, 1861) adopted a narrower view of *Biddulphia*, separating *Cerataulus* on the convexity of its valve and *Zygoceros* on the fact that its frustules were free and not united by the corners into zig-zag chains. Like Kützing (1844) earlier, he placed these genera, along with *Porpeia* Bail. ex Ralfs, *Hemiaulus* Ehrenb., *Isthmia* Ag. and *Hydrosera* Wallich, all with limits similar to those recognized today, in one family, the Biddulphieae, separate from the multiangular genera *Triceratium*, *Amphitetras* and *Amphipentas*, with which, in his family Anguliferae, he included *Euodia* Bail. ex Ralfs and *Hemidiscus* G. C. Wallich. These two latter genera are, in fact, synonymous and not at all related to the genera we are considering but instead close to *Actinocyclus* Ehrenb.; a small number of biddulphioid species have, however, been referred to *Euodia*.

H. L. Smith (1872) took a much wider view of the limits of genera than had Ralfs. He united the Biddulphieae, Anguliferae and Terpsinoeae of the latter in a single family, Biddulphieae, containing only six genera. He did not accept the difference in number of poles or angles as a basis for generic distinction but included *Triceratium*, *Amphitetras* and *Amphipentas* in *Biddulphia*, in which he also placed *Cerataulus*, *Odontella* and *Zygoceros*.

Grunow (1884), like H. L. Smith, did not separate bipolar from multiangular species but divided H. L. Smith's genus *Biddulphia* into two, *Biddulphia* and *Odontella*. In *Biddulphia* he retained the forms with rounded processes bearing punctation similar to, although smaller than, that on the rest of the valve and not sharply separated from it. In *Odontella* he placed those with truncate processes with an area at the tip sharply delimited by a double border. This border is in fact not double; the two lines that appear under the microscope are the inner and outer edges of a thickened rim. Grunow further suggested that a group of species with a sharp dividing line at the junction of the valve surface and the mantle on which there are well-developed spines might be separated from *Odontella* as a separate genus *Denticella*.

De Toni (1890) accepted this suggestion of Grunow's as far as bipolar forms were concerned but continued to distinguish genera on the basis of the number of angles. He grouped the bipolar species into four genera:

> *Biddulphia* with processes rounded above with puncta becoming smaller towards the tip;
> *Odontella* with abruptly truncate processes;
> *Denticella* with awl-shaped setae or spines on the central part of the valve;
> *Cerataulus* with discoid valves with processes alternating with horn-like spines.

The recognition of *Denticella* as a genus separate from *Odontella* is nomenclaturally illegitimate; the two genera have the same type species. De Toni maintained *Amphitetras* and *Amphipentas* for species with four-angled and five-angled valves respectively, noting that the two genera were scarcely distinct, used *Triceratium*, quite incorrectly, for those whose valves have more than five angles and *Trigonium* Cleve for those with triangular valves. *Trigonium* is a genus proposed by Cleve (1868) for diatoms with triangular valves and no definite processes at the angles. His separation of it from *Triceratium* was the first suggestion that the multangular species should be divided on the basis of their structure.

De Toni (1894), in his later and more extended treatment, maintained the same generic limits but transferred *Cerataulus* to the *Eupodiscaceae*; he also coined the new generic name *Nothoceratium* De Toni for the species whose valves had six or more angles and used *Triceratium* instead of *Trigonium* for those whose valves had three angles. It is noteworthy that, whilst he divided the bipolar forms into four genera, he included in a single genus triangular forms having quite as wide a range of structure.

Two classifications of the diatoms appeared in 1896. In both of these wider generic limits in the group with which we are concerned were adopted. Van Heurck (1896) took as wide a view of *Biddulphia* as had H. L. Smith, including within it the species placed by De Toni in *Biddulphia*, *Odontella*, *Denticella*, *Cerataulus*, *Triceratium*, *Amphitetras*, *Amphipentas* and *Nothoceratium*, and also the angular species of *Stictodiscus* Grev. Schütt (1896) on the other hand separated the bipolar forms from the multangular ones in different sub-tribes, but united the four of De Toni's bipolar genera listed in the previous sentence in one genus,

Biddulphia, and the multiangular species, including multiangular *Stictodiscus*, also in a single genus, *Triceratium*.

The next treatment of the diatoms with which we are concerned to differ from those already put forward is Mann's (1907). He recognized two genera, *Biddulphia*, with definite processes at the poles or angles, and *Trigonium* without such definite processes, combining in both bipolar and multiangular forms, but with very few of the former in *Trigonium*.

The brothers Peragallo (H. and M. Peragallo, 1908), on the other hand, separated the species with rounded processes or scarcely elevated angles covered with puncta similar to but smaller than those of the valve from those with truncate processes with the apex clearly delimited from the rest of the valve. They also kept bipolar and multiangular species in different genera, recognizing four altogether, *Biddulphia* and *Trigonium* in the first group, *Odontella* and *Triceratium* in the second, with *Cerataulus* kept separate and placed in the same family as *Auliscus* Ehrenb., i.e. the Eupodiscaceae of De Toni.

Meanwhile Gran (1905) had put forward the view that torsion about the pervalvar axis was the distinguishing character of the genus *Cerataulus*. Hustedt (1930) followed him in this, as have all subsequent authors. The species without this torsion were placed in two genera, *Biddulphia* and *Triceratium*, by Hustedt who argued that the difference between bipolar and multiangular symmetry was of fundamental importance. It is, of course, true that the frustules of bipolar forms have three planes of symmetry at right angles, no one of which is equivalent to another whilst multiangular forms have as many equivalent planes of symmetry at right angles to the valvar plane as they have angles. The adoption of a terminology for the symmetry of the frustules that emphasizes this distinction rather than one which speaks simply of bipolar, tripolar, quadripolar, etc., valves gives this argument a spurious force. This delimitation of the genera, nevertheless, has been followed by most authors since 1930 and can be regarded as the currently orthodox view.

Hendey (1937, 1964) is the only author since 1930 to have put forward an alternative classification of the species with which we are concerned. He combines in one genus, *Biddulphia*, most of the species included by Hustedt in both *Biddulphia* and *Triceratium* but separates two groups of multiangular species in genera with comparatively narrow limits, *Triceratium* and *Trigonium*; he restricts *Triceratium* to species whose valves have three or more angles, alveoli in straight lines parallel to the margins, and stout cornutate processes, and *Trigonium* to species whose valves have three or more rounded corners without elevated processes and radially arranged alveoli that grade into areas of much finer alveoli towards the corners.

From this brief survey two points emerge: (1) there has been great diversity of opinion on the question of generic limits in the Biddulphiaceae and no generally accepted consensus has emerged; (2) many authors have attached considerable importance to the distinction between bipolar and multiangular forms, and those who have done so have been more ready to divide the bipolar forms into separate genera on the basis of structure than they have been to recognize similar divisions among the multiangular ones, Hendey being the one exception.

MATERIAL AND METHODS

The specimens studied have come from material from various sources in the collections of the British Museum (Natural History). The localities are given in the account of the observations on the various species. The gatherings have all been subjected to chemical treatment to clean the silica of organic matter. The details of this have not normally been recorded but in almost all cases it will have been the treatment most frequently used by workers on diatoms, boiling in concentrated sulphuric acid followed by the cautious addition to the boiling acid of crystals of potassium chlorate. Individual specimens from strews of cleaned material dried onto mica slips have been transferred to specimen stubs and affixed by double-sided "Sellotape", gum arabic or gum tragacanth, coated with gold or gold-palladium and examined in a Cambridge Instrument Co. Stereoscan scanning electron microscope at 20 kV accelerating voltage.

OBSERVATIONS

Before describing the structure of the species that we have examined, there is one point of terminology that needs explanation. The valves of diatoms are perforate, the perforations being the markings that, along with their shape, provide the characters by which species have been recognized under the light microscope. The transmission electron microscope has provided considerable additional information about the structure of these perforations. Various elaborate schemes of classing the structure found have been put forward (e.g. Hendey, 1959; Geissler et al., 1961), but in the light of the additional information that the scanning electron microscope has given us, they seem both inadequate and unduly complex. We plan to discuss this subject in more detail and to put forward a scheme covering the diatoms as a whole in a later paper. In the species with which we are concerned here, however, we recognize only two types of structure of valve, porose and alveolate.

By a pore we mean a hole through the valve with a uniform bore throughout or narrowing towards the centre of the thickness of the valve and then expanding again (see Plate V, Fig. 3). This may or may not have a membrane, more or less elaborately structured, across it at any level. An alveolus, on the other hand, is a chamber in the thickness of the valve opening to the inside or outside of the frustule by a pore or slit of smaller diameter than the lumen of the alveolus; (Plate III, Fig. 4; Plate IV, Fig. 2); at the surface of the valve opposite that on which the pore or slit opens there is usually, perhaps always, a membrane of more or less elaborate structure. Pores are normally circular and well separated. Alveoli are mostly polygonal and thin-walled.

Throughout the remainder of this paper we use for the species being discussed the names in current use by those who accept Hustedt's (1930) views on the limits of the genera concerned, even although, as will become apparent, we disagree in many places with those limits.

1. Biddulphia biddulphiana (*J. E. Smith*) *Boyer*

This species is the type of the genus *Biddulphia* Gray. The material examined is a recent gathering from Cape Agulhas, South Africa. The valve (Plate I, Fig. 1) has a deep and vertical mantle, and a crenate margin with slight furrows running across the valve from the inner points of the crenations. There are deep transverse ribs projecting into the interior below these furrows (Plate I, Fig. 4). The processes at the two poles are large and rounded (Plate, I, Fig. 2). On the main body of the valve there are pores arranged in more or less radial rows. These are about 2μm in diameter and 3μm from centre to centre (3–3·5 in 10μm); across them there is an elaborately structured plate (Plate I, Fig. 3). Small simple spines are present on the margins of many of them. The processes are covered by much smaller pores which also have plates within them, and there is a gradation in size between the pores of the valve and those on the processes, which are about $0·2\mu$m in diameter and $0·5\mu$m centre to centre (20 in 10μm).

There are other species of *Biddulphia*, e.g. *Biddulphia tuomeyi*, that are very similar to *Biddulphia biddulphiana* but have more strongly developed spines, including, in the central portion of the valve, a few large hollow spines opening on the inside by slits across rounded projections similar to those in *Triceratium spinosum* (see below).

2. Triceratium stokesianum *Grev.*

The specimen examined comes from the Miocene of Szurdokpuspoki, Hungary (Chenevière, 1933, 1934). It is triangular (Plate I, Fig. 5) but resembles *Biddulphia biddulphiana* in having a deep vertical valve mantle, internal ribs, pores with a complex membrane across them and processes with similar but smaller pores with a gradation between those of the general surface and those of the process (Plate I, Fig. 7). The pores of the main part of the valve are $2·5\mu$m in diameter and about 5μm from centre to centre (2 in 10μm); those on the summit of the processes are about 0·6μm in diameter and about 0·8μm centre to centre (12 in 10μm). We have seen a number of other species of *Triceratium* which also possess these characters, although they differ in the dimensions and spacing of the pores. One feature special to *Triceratium stokesianum*, as far as our current investigations go, is that it bears not spines but small spherical excrescences on the surface with a ring of rather larger ones in the centre with a perforation through them that opens on the inner side by a slit across a rounded projection (Plate I, Fig. 6).

3. Triceratium formosum *Brightw.*

The specimens examined come from a recent gathering from Paama, New Hebrides. This species (Plate II, Fig. 1) resembles *Biddulphia biddulphiana* and *Triceratium stokesianum* in having a deep vertical mantle, but it has no internal ribs and no spines. The rounded corners of the valve are scarcely raised (Plate II, Fig. 2) but the markings over them are much finer than those on the remainder of the valve. The main difference between this species and the two previously described is in the structure of the valve; this is alveolate, not porose (Plate II, Figs 3, 5). The alveoli open to the interior of the frustule by circular pores; on the outside they are capped by a domed membrane with five or six partially occluded holes around the periphery of the dome. The alveoli are about 2·5μm centre to centre (4 in 10μm). Towards the corners they become much smaller, the membrane over them becomes less domed, and the number of holes per alveolus is reduced, until it is no more than two, that are semicircular and without any partially occluding membranes across them (Plate II, Figs 4, 6). On the summits of the processes the alveoli are 0·65μm centre to centre (16 in 10μm).

This species is one of those that Hendey would include in the genus *Trigonium* Cleve. We have examined other species within his concept of the genus, including its type, *T. arcticum* (Brightw.) Cleve, but have not obtained satisfactory pictures of them as yet. Our observations show, however, that only minor differences in the size and arrangement of the alveoli and the structure of the membrane on their outer side distinguishes them.

4. Triceratium spinosum *Bail.*

The specimens examined are from recent material from Tampa, Florida. As in *Triceratium formosum*, the structure of the valves is alveolate, not porose (Plate III, Figs 3, 4). The outer membrane of each alveolus is slightly sunk and has a number of perforations. The alveoli are about 3μm centre to centre (about 3 in 10μm). On the surface of the valve, arising from the slight ridges that separate the membranes covering the alveoli, there are scattered spines; these spines are buttressed at the base and are not pointed but are instead blunt or lobed. Many small granules also occur on the same ridges. There are three stout hollow spines, one proximal to each corner, but somewhat offset from the line joining corner and centre (Plate III, Fig. 1); the tube through the centre of the spines opens to the interior of the frustule by a slit across a rounded protuberance (Plate III, Fig. 4).

The characters so far mentioned, except for minor differences in detail, are also possessed by one or other of the species so far discussed. There are two respects, however, in which *Triceratium spinosum* differs markedly from all of them. The most striking is the process. There is no gradual transition between the structure of the main body of the valve and that of the process. Instead, the alveoli extend up the side of the process without modification and at the tip there is a solid ring of silica within which there is a plate $8 \mu m$ in diameter penetrated by pores which seem to have no membrane across them (Plate II, Fig. 2). These pores are closely packed; their diameter is about $0.15 \mu m$ and they are about $0.25 \mu m$ centre to centre (40 in $10 \mu m$). The other point of obvious difference is that the mantle of the valve is not vertical but indented, curving out again to the margin (Plate III, Fig. 1).

5. Biddulphia rhombus (*Ehrenb.*) *W. Smith*

The specimens of this species that were examined came from a recent gathering from Tampa, Florida. This is the type species of *Zygoceros* Ehrenb. It is a bipolar species (Plate III, Fig. 5) with an alveolate valve structure, an indented valve mantle and processes resembling those of *Triceratium spinosum* in structure (Plate III, Fig. 6). The alveoli are radially arranged and are approximately $1.3 \mu m$ centre to centre (7.5 in $10 \mu m$). The membrane on the outer surface has rows of small holes forming a polygonal pattern, with 1–3 granules in the centre of each polygon. The polygonal pattern is, however, disturbed by the many

---

Plate I

Figs 1–4, *Biddulphia biddulphiana* (J. E. Smith) Boyer.
Fig. 1. Whole valve, outer view, $\times 400$.
Fig. 2. Apical part of valve with break showing porose structure, outer view, $\times 1500$.
Fig. 3. Detail of edge of process, outer view, $\times 7500$.
Fig. 4. Interior of apical part of valve showing transverse rib, $\times 2750$.
Figs 5–7. *Triceratium stokesianum* Grev.
Fig. 5. Whole valve, outer view, $\times 300$.
Fig. 6. Central portion of valve, outer view, $\times 2500$.
Fig. 7. Process, outer view, $\times 2500$.

9. *Generic Limits in Biddulphiaceae as Indicated by the SEM*   163

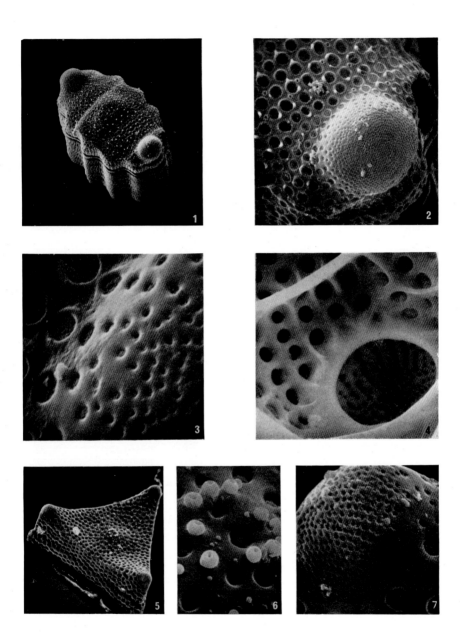

buttressed spines with blunt lobed tips that occur on the valve surface (Plate III, Fig. 7). There are in addition three of four large hollow spines close to each of the two lateral margins of the valve and one close to each process but offset from the apical axis of the valve (Plate III, Figs 5, 6). These spines open to the interior of the valve by a slit across a rounded protuberance. The perforate plate at the apex of the process is much smaller than in *Triceratium spinosum*, only about 3μm in diameter, and the perforations are somewhat closer, $0\cdot15\mu$m centre to centre (about 65 in 10μm).

6. Cerataulus turgidus (*Ehrenb.*) *Ehrenb.*

The material of this species, which is the type species of the genus *Cerataulus* Ehrenb., came from a recent gathering from the Bonny River, Nigeria. Its frustule has a slight twist and its two processes are offset from the apical axis of the valve, which is alveolate (Plate IV, Figs 1, 2). The alveoli are about $0\cdot8\mu$m centre to centre (12 in 10μm) and their outer membrane is granular (Plate IV, Fig. 2). There are many spines on the surface and these also are buttressed. The mantle is indented. There is a group of two or three large hollow spines on either side of the valve nearer the margin than the centre and diagonally opposite each other (Plate IV, Fig. 1); on the inside of the valve the tube through the spine opens through a slit in a rounded protuberance. The perforated plate at the apex of the processes is about $7\cdot5\mu$m in diameter and the perforations themselves are about $0\cdot225\mu$m centre to centre (45 in 10μm).

7. Triceratium antediluvianum (*Ehrenb.*) *Grun.*

Miocene material from Carro del Pontón, Morón, Spain, is the source of the specimens of this species that were examined. This is the type species of the genus *Amphitetras* Ehrenb. The valves, which are quadrangular (Plate V, Fig. 1), have indented mantles and are porose, each pore normally having across it a finely perforate plate with a ring of larger perforations around its outer margin; the

---

Plate II

Figs 1–6. *Triceratium formosum* Brightw.
Fig. 1. Whole valve, outer view, $\times 500$.
Fig. 2. Whole frustule, $\times 500$.
Fig. 3. Detail of central part of valve, outer view, $\times 5000$.
Fig. 4. Angle of valve, outer view, $\times 2500$.
Fig. 5. Detail of valve, inner view showing alveolate structure, $\times 5000$.
Fig. 6. Angle of valve, inner view, $\times 2500$.

9. Generic Limits in Biddulphiaceae as Indicated by the SEM

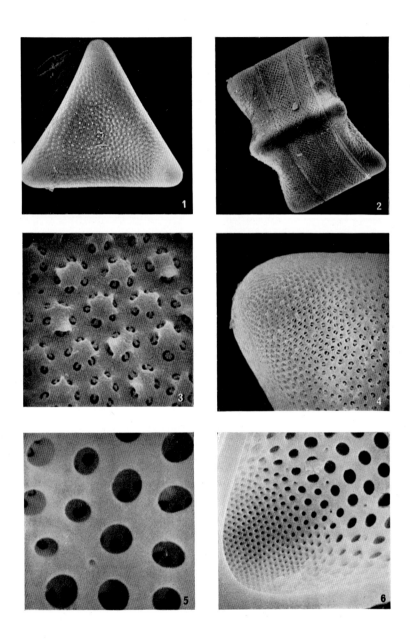

plate is somewhat sunk below the surface of the valve and is slightly convex outwards (Plate V, Figs 2, 3). Occasional pores, however, have across them a membrane flush with the surface of the valve that has in it three or four kidney-shaped openings surrounding a granule or small perforation. The pores are ca 2μm in diameter and 3–4μm centre to centre (2·5–3 in 10μm). There is a mesh-work of low and very narrow ridges on the surface of the valve with each mesh of the ridge normally enclosing only a single pore (Plate V, Fig. 2). On these ridges there are a number of very small granules, and scattered over the surface of the valve are short blunt buttressed spines. On each valve there are four hollow spines, opposite the centre of each side and about one-third of the distance from the margin of the valve to the centre; these open to the interior through slits across rounded projections. At each apex there is a large circular plate perforated by pores and surrounded by a solid rim; the diameter of the plates tends to vary from one corner to another and is in the range 10–14μm. The pores themselves are about 0·22μm centre to centre (45 in 10μm). The break in the specimen illustrated as Plate V, Fig. 3 passes through the plate and shows that the perforations in it are simple pores.

8. Triceratium pentacrinus (*Ehrenb.*) G. C. Wallich

The specimens studied come from a recent gathering made at Paama, New Hebrides. This is the type species of the genus *Amphipentas* Ehrenb. The valves, which may have four or five angles, have indented mantles (Plate IV, Fig. 3). The prominent feature of this species is the network of strong ridges over the surface of the valve. This network is somewhat irregular and the meshes are largest in the depressed central part of the valve and in the depression between the raised ring around the central depression and the angles. The valve is porose

---

Plate III

Figs 1–4. *Triceratium spinosum* Bail.
Fig. 1. Whole valve, outer view, ×300.
Fig. 2. Process, outer view, ×3000.
Fig. 3. Central part of valve showing hollow spine, outer view, ×1250.
Fig. 4. Central part of valve showing openings of hollow spines into interior and broken edge showing alveolate structure, inner view, ×1250.
Figs 5–7. *Biddulphia rhombus* (Ehrenb.) W. Smith.
Fig. 5. Whole valve, outer view, ×350.
Fig. 6. Process and hollow spine, outer view, ×7000.
Fig. 7. Surface of valve showing hollow spine and solid spines, oblique outer view, ×4000.

9. Generic Limits in Biddulphiaceae as Indicated by the SEM 167

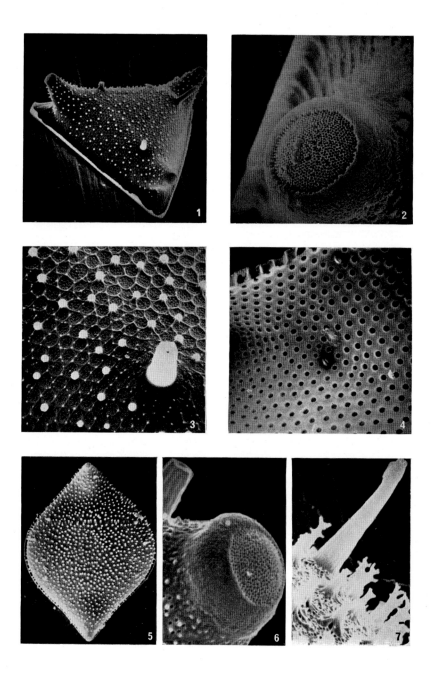

and there are a number of pores within each mesh of the network of ridges (Plate IV, Fig. 4). The pores are about 1μm in diameter and about 2μm centre to centre (5 in 10μm). They have a domed plate across them level with the outer surface of the valve. This has a structure similar to the plate in *Triceratium antediluvianum* but with a number of small granules on it that do not seem to be present in that species. There are no solid spines on the surface of the valve but there is a hollow spine opposite the centre of each side in the same position as the comparable spine in *Triceratium antediluvianum* and it opens in the same way on the inner side of the valve. The processes at the angles have at their apices a plate perforated by pores and surrounded by a solid rim; the diameter of the plate is *ca* 5μm and the pores are about 0·2μm centre to centre (50 in 10μm).

9. Biddulphia reticulata *Rop*.

The specimens examined are from recent material from Thursday Island, Queensland, Australia. Some are bipolar (Plate IV, Fig. 5), others triangular. The valves, which have an indented mantle, have a very similar structure to those of *Triceratium pentacrinus*. The meshes of the network of ridges are more even in size than those of that species but their arrangement is irregular. The crests of the ridges are somewhat expanded sideways by flanges so that there are rounded openings at this level into irregularly polygonal chambers beneath (Plate IV, Fig. 6). The valve proper is porose with domed membranes across the pores at their outer end; the pores are of variable size from 0·3 to 0·7μm in diameter and *ca* 1·0μm centre to centre (10 in 10μm). On the crests of the ridges there are buttressed spines of varying sizes and, in most specimens, there is a hollow spine a short distance proximal to each angle but offset from the line joining the angle to the centre of the valve; this opens on the inner side, like those in other species, through a slit in a projection. These spines are apparently sometimes absent; there

---

Plate IV

Figs 1–2. *Cerataulus turgidus* (Ehrenb.) Ehrenb.
Fig. 1. Whole valve, outer view, ×750.
Fig. 2. Broken edge of valve showing alveolar structure, ×6000.
Figs 3–4. *Triceratium pentacrinus* (Ehrenb.) G. C. Wallich.
Fig. 3. Whole valve, outer view, ×600.
Fig. 4. Angle of valve showing process, ribs and pores, ×3500.
Figs 5–6. *Biddulphia reticulate* Rop.
Fig. 5. Whole valve, outer view, ×750.
Fig. 6. Angle of valve showing process, chambers and pores, ×3500.

9. Generic Limits in Biddulphiaceae as Indicated by the SEM   169

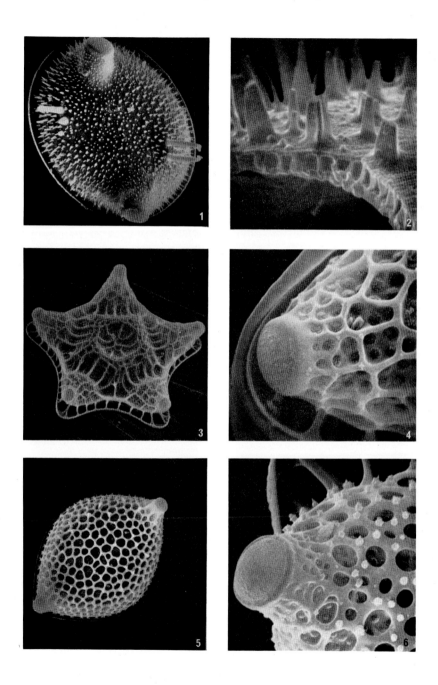

is no trace of them on the specimen figured as Plate IV, Fig. 5. The processes at the poles are capped by a perforate plate surrounded by a solid rim. The diameter of the plate is *ca* 5μm and the perforations are rather less than 0·2μm centre to centre (55 in 10μm).

10. Triceratium favus Ehrenb.

The material illustrated came from recent gatherings from Bonny River, Nigeria, Tampa, Florida, and the estuary of the River Thames, England. We have, however, also examined material from the Upper Eocene of Oamaru, New Zealand, which shows no appreciable differences. The valve has on its outer side chambers similar to those of *Biddulphia reticulata* except that they are hexagonal and regularly arranged (Plate V, Fig. 4). There are small spines at the corners of these and on the floor of them there are pores arranged in radiating rows (Plate V, Fig. 6). These pores are 0·3μm in diameter and 0·4 to 0·5μm centre to centre (20–25 in 10μm) along the rows. They are capped by a domed membrane similar to that in *Biddulphia reticulata* (Plate V, Fig. 7). These membranes across the pores have not previously been seen and the structure of this diatom has been interpreted in the past as being alveolate with the alveoli opening outwards by a pore and closed on the inner side by a perforate membrane, i.e. like *Triceratium spinosum* inside out. The series *Triceratium pentacrinus*, *Biddulphia rhombus*, *Triceratium favus*, however, shows that this interpretation is incorrect and that we are dealing here with a porose valve with chamber-forming ridges on its outer side.

Each process is capped by a perforate plate 3·5–4·0μm in diameter, the perforations themselves being *ca* 0·25μm centre to centre (40 in 10μm) (Plate V, Fig. 5). The pores of the valve do not approach the perforate plate as closely as

---

Plate V

Figs 1–3. *Triceratium antediluvianum* (Ehrenb.) Grun.
Fig. 1. Whole valve, outer view, ×500.
Fig. 2. Process, outer view, ×1500.
Fig. 3. Broken edge of valve showing porose structure of valve and structure of perforate plate of process at upper left, ×2500.
Figs 4–7. *Triceratium favus* Ehrenb.
Fig. 4. Whole valve, outer view, ×500, specimen from Bonny River, Nigeria.
Fig. 5. Process, outer view, ×3500, specimen from Tampa, Florida.
Fig. 6. Central part of valve, inner view, ×2500, specimen from Bonny River, Nigeria.
Fig. 7. Chamber showing membranes over pores, outer view, ×7000, specimen from estuary of River Thames, England.

9. Generic Limits in Biddulphiaceae as Indicated by the SEM 171

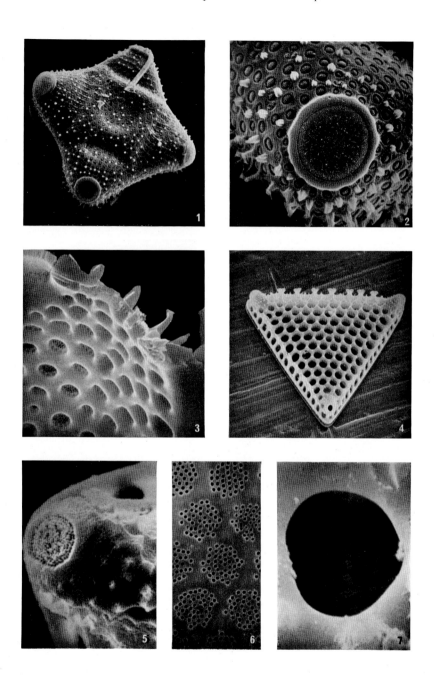

in the previous two species and the greater part of the raised process is solid silica and bears many small buttressed spines.

Along the ridge that runs from process to process, immediately proximal to the rows of chambers along the margin of the valve, there is a row of hollow spines that are expanded sideways into wings. These open through slits in rounded projections on the inner side of the valve.

Since the valve surface proper is the outer surface of the layer that is the floor of the chambers, it is clear that in this species also the valve mantle is not vertical but slopes outwards to the margin.

DISCUSSION AND CONCLUSIONS

The species described in the previous section fall obviously into four groups:
- valves porose; markings on valve grading into those on processes or angles;
- valves alveolate; markings on valve grading into those on processes or angles;
- valves porose; processes or angles with finely perforate plate surrounded by a solid rim;
- valves alveolate; processes or angles with finely perforate plate surrounded by a solid rim;

In each of these groups there are bipolar and multiangular forms, and, in all but the first, ones so similar that authors such as Hustedt, who normally divide the bipolar and multiangular forms into separate genera, treat them as belonging to the same species. It therefore seems to us that any separation on these grounds is unjustified.

The four groups that we have listed above are in our view so distinct from one another that there can be no question but that they consist of separate genera. We have now to consider whether each group contains one genus only, or more than one.

As we indicated above, we have examined a number of species with porose valves and pores on the processes that grade into those on the main part of the valve. Nineteen such species have complete or partial internal ribs; these include not only *Biddulphia biddulphiana*, *B. tuomeyi* and *Triceratium stokesianum* described in the previous section but also *Triceratium castelliferum* Grun., *T. crenulatum* Grove & Sturt, *T. flos* Ehrenb., *T. glandiferum* Grun., *T. majus* (Grove & Sturt) Grove & Sturt ex Grun., *T. morlandii* Grove & Sturt, *T. nova-zealandicum* (Grove & Sturt) Grun. and *T. polycistinorum* Pant., as well as others that await identification. It is only differences in detail that separate species from species amongst these, and our tentative conclusion is that they all belong to one genus, *Biddulphia* Gray.

Among the species we have examined there are also seven that differ from these last in having no internal ribs. These are *Triceratium fractum* Walk. and Chase, *T. inelegans* Grev., *T. plano-concavum* Brun, *T. reticulum* Ehrenb., *T. rugosum* Grove & Sturt, *T. shadboltianum* Grev. and a species so far unidentified from the Eocene of Kamychin, U.S.S.R. We are uncertain as yet about the proper generic disposition of these; some at least, possibly all, should probably be referred to *Biddulphia*, but there may be one or more other genera represented, which, or one of which, is close to the multiangular species of *Stictodiscus* Grev.

Of the species with an alveolate valve and with the alveoli of the central part of the valve grading into those on the processes we have examined three species, *Triceratium formosum* described above, *T. arcticum* Brightw. from both a recent gathering and from the Miocene of California, and *T. montereyi* Brightw. from the Miocene of California and the Upper Eocene of Oamaru, New Zealand. These, as already noted, are very similar and the only differences are in contour, closeness of packing of the alveoli in the centre of the valve, and detail of the outer membrane of the alveoli. We therefore believe that they constitute a single genus, *Trigonium* Cleve, with limits similar to those suggested by Hendey (1937, 1964). The appearance under the scanning electron microscope of a specimen of *T. margaritaceum* Ralfs from the Miocene of Maryland, U.S.A., the fine structure of which seem to have been considerably modified by fossilization, suggests that that species may belong here.

We have also obtained pictures of a triangular species from the Eocene of Kamychin, U.S.S.R., which has an alveolate valve with the structure of the outer membrane similar to that found in *Trigonium*. The alveoli scarcely diminish in size towards the angles, however, and at each angle there is an elongate pore on the outer surface that leads to an elliptical tube projecting into the cell with a slit across its tip. Similar structures occur in such genera as *Actinocyclus* Ehrenb. and *Coscinodiscus* Ehrenb., but we have not seen them in any other species that might belong to the Biddulphiaceae. The generic position of this species remains obscure.

The group of species with alveolar valves and processes terminating in a finely perforate plate also shows considerable uniformity in all but minor detail. Of such forms we have examined not only *Biddulphia rhombus*, *Ceratalus turgidus* and *Triceratium spinosum* described in the previous section of this paper but also *Biddulphia mobiliensis* (Bail.) Grun, *Biddulphia armata* Leud.–Fortm. and a recent triangular species from Java as yet unidentified. The only problem that they present as regards generic disposition is whether the torsion and the off-setting of the processes from the major axis of the valve that characterize *Ceratalus turgidus* and the other species currently placed in that genus is a sufficient basis for generic distinction. For the present we consider that it will probably be best to maintain this distinction and to treat this group as containing two genera, *Zygoceros* Ehrenb. without torsion and *Cerataulus* Ehrenb., with torsion.

The problem of generic limits in the group with porose valves and perforate plates on the processes is, however, more difficult. We have found this type of structure not only in *Triceratium antediluvianum*, *T. pentacrinus*, *T. favus* and *Biddulphia reticulata* described above but also in *B. parallela* Castr., *B. polyacanthos* Brun and *B. zanzibarica* A. Schmidt, none of which possess ridges on the outer surface, and also a bipolar form of *T. dubium* Brightw., which has a fine structure very similar to that of *B. reticulata*, and *Triceratium secedens* A. Schmidt, which closely resembles *T. favus*. *B. parallela* and *B. zanzibarica* are included by Hustedt (1930) in *B. aurita* (Lyngb.) Bréb. but we believe them to be separate species. *B. aurita*, which is the type of *Odontella* Ag., shows no sign of spines, other than two or three hollow ones near the centre of the valve, when examined under the light microscope; it does appear, however, to have porose valves and processes with porose plates but this will need to be confirmed under the scanning electron microscope.

The range of structure in this group of species is very considerable and a genus that included both *Triceratium antediluvianum* and *T. favus* would be very heterogeneous. Nevertheless it is by no means easy to see where distinctions might be drawn. The species such as *Triceratium favus* that have a comparatively regular arrangement of chambers and rows of hollow spines running from process to process form a compact unit. A genus delimited by the possession of these characters would come close to Hendey's (1937, 1964) concept of *Triceratium* Ehrenb., but would exclude *Biddulphia reticulata*, and *Triceratium dubium*, which are very similar except in the disposition of their hollow spines. These would be included with *T. favus* and its allies if the criterion adopted for delimiting a genus were the presence of chambers on the outer side of the valve opening outwards through large pores formed by flanges on the tops of the ridges separating the chambers. This, however, would exclude *T. pentacrinus*, which has its ridges not expanded at the summit. This species would be included with all the previous ones if a separation were made on the basis of the possession of a meshwork of well-developed ridges on the outer side of the valve, each mesh enclosing more than one pore. To make a division here would separate *T. pentacrinus* from *T. antediluvianum*, where the ridges are only slightly developed and have, for the most part, only a single pore in each mesh. The two are nevertheless very alike in all other respects. Continuing this line of argument further, there is nothing other than the placing of the hollow spines, which is identical in *T. pentacrinus* and *T. antediluvianum*, that might serve as a generic criterion to separate the latter from *Biddulphia parallela*, *B. polyacanthos* and *B. zanzibarica*, between which there are no differences that we can detect that are of more than specific value.

Thus, whilst the dissimilarities between *Triceratium favus* and *T. antediluvianum* seem too great to permit their being regarded as belonging to a single genus, there seems no obvious point between them at which to draw the line. We must therefore leave it for the present as an open question whether this group of species constitutes a single genus, for which the correct name would probably be *Odontella* Ag., or more than one, with *Triceratium* Ehrenb. and perhaps also *Amphitetras* Ehrenb. or *Amphipentas* Ehrenb. separated from *Odontella*. After examination of further species within the group it may be possible to resolve this question, perhaps by means of the techniques of numerical taxonomy.

Another question that raises itself at this stage is the way in which the genera that we have tentatively delimited should be classified into higher groups. At first sight there would seem to be two possibilities, to separate those with porose valves from those with alveolar valves, or alternatively to distinguish between those with perforate plates at the apex of the processes from those with the markings on the processes grading into those of the valve. If the distinction is made on the former basis, however, no other character that we have noticed would follow the same pattern. On the other hand, if the latter division is made, the groups would be distinguished by the characteristics set out in tabular form below.

Processes with delimited perforate plate	Processes not sharply delimited from remainder of valve
Mantle turned outwards to margin of valve	Mantle vertical
Ribs or ridges, when present, external	Ribs or ridges, when present, internal
Hollow spines, when present, close to processes or margins, or in two opposite groups of two or three.	Hollow spines, when present, in an irregular group in central part of valve.

If a division is made on this basis, such genera as *Eunotogramma* Weisse, *Leudugeria* Temp. ex Van Heurck, and *Terpsinoe* Ehrenb., which we have examined under the scanning electron microscope, would be associated with *Biddulphia* and *Trigonium*, and the light microscope suggests that here too would belong

Anaulus Ehrenb., *Hydrosera* G. C. Wallich, and *Porpeia* Bail. ex Ralfs. *Hemiaulus* Ehrenb. and *Trinacria* Heib. are sufficiently similar in their general fine structure to these, although having distinctive structures associated with their processes, to make it reasonable to include them in a family Biddulphiaceae with subfamilies Biddulphioideae and Hemiaulioideae.

Zygoceros, Cerataulus, Odontella, if we interpret its identity aright, and *Triceratium* together with any other genera that may be recognized in this group, would not belong to this family. The so-called ocelli of *Auliscus* Ehrenb., *Eupodiscus* Bail, *Pseudauliscus* Leud.–Fortm. and *Rattrayella* De Toni are perforate plates of the same type as those on the processes of the above genera. To include these latter in a family Eupodiscaceae in the sense of Hustedt (1930), but with *Aulacodiscus* Ehrenb. excluded (see Ross and Sims, 1970), would therefore seem to us the correct disposition to make.

REFERENCES

CHENEVIÈRE, E. (1933, 1934). Note sur le depôt de terre a diatomées fossiles (Miocéne supérieur) récemment découvert prés de Szurdokpuspoki. *Földt. Közl.* **63**, 216–218. Reimpr. in *Bull. Soc. fr. Microsc.* **3**, 33–36.

CLEVE, P. T. (1868). Diatomaceer från Spetsbergen. *Ofvers. K. VetenskAkad. Förh. Stockh.* **24**, 661–669.

DE TONI, G. B. (1890). Osservazioni sulla tassonomia delle Bacillariee (Diatomee) seguite da un prospetto dei generi delle medesime. *Notarisia* **5**, 885–922.

DE TONI, G. B. (1894). *Sylloge Algarum* **2**(3). Patavii, pp. 819–1426.

GEISSLER, U., GERLOFF, J., HELMCKE, J.-G., KRIEGER, W. and REIMANN, B. (1961). *Diatomeenschalen im Elecktronenmikroskopischen Bild*, Vol. **3**, 44 pp., taf. 201–300. Cramer, Weinheim.

GRAN, H. H. (1905). Diatomeen. *In:* K. Brandt and C. Apstein, *Nordisches Plankton* **19**, 1–146.

GRUNOW, A. (1884). Die Diatomeen von Franz Josefs-Land. *Denkschr. Akad. Wiss. Wien* **48**, 53–60, taf. I–V.

HENDEY, N. I. (1937). The plankton diatoms of the southern seas. *"Discovery" Rep.* **16**, 151–364, pl. VI–XIII.

HENDEY, N. I. (1959). The structure of the diatom cell wall as revealed by the electron microscope. *J. Quekett microsc. Club* ser. 4, **5**, 147–175.

HENDEY, N. I. (1964). An introductory account of the smaller algae of British coastal waters. Part V: Bacillariophyceae (Diatoms). *Fishery Invest.* ser. 4, **5**, 1–317, pl. I–XLV.

HUSTEDT, F. (1930). Die Kieselalgen Deutschlands, Österreichs und der Schweiz mit Berucksichtigung der übrigen Länder Europas sowie der angrenzenden Meeresgebiete. *In* L. Rabenhorst: "Kryptogamen-Flora von Deutschland, Österreich und der Schweiz", Vol. **7**, pp. 609–920, Leif. 4, 5, Akademische Verlag, Leipzig.

KÜTZING, F. T. (1844). "Die Kieselschaligen Bacillarien oder Diatomeen". 152 pp., 30 plates, Köhne, Nordhausen.

MANN, A. (1907). Report on the diatoms of the Albatross voyages in the Pacific Ocean, 1888–1904. *Contr. U.S. natn. Herb.* **10,** i–viii, 221–442, pl. XLIV–LIV.

PERAGALLO, H. and M. PERAGALLO (1908). "Diatomées Marines de France", Pt. 3, pp. 365–491, Tempère, Grez-sur-Loing.

PRITCHARD, A. (1861). "A History of Infusoria including the Desmidiaceae and Diatomaceae, British and Foreign". 4th Ed. 968 pp., 39 plates, Whittaker, London.

ROPER, F. C. S. (1859). On the genus Biddulphia and its affinities. *Trans. microsc. Soc. Lond.* new ser., **7,** 1–24, pl. I–II.

ROSS, R. (1963). Ultrastructure research as an aid in the classification of Diatoms. *Ann. N.Y. Acad. Sci.* **108,** 396–411.

ROSS, R. and SIMS, P. A. (1970). Studies of *Aulacodiscus* with the scanning electron microscope. *Beih. Nova Hedwigia* **31,** 49–88, pl. 1–13.

SCHÜTT, F. (1896). Bacillariales. *In:* A. Engler and K. Prantl, *Naturl. PflFam.,* **1(1b),** 31–150.

SMITH, H. L. (1972). Conspectus of the families and genera of the Diatomaceae. *Lens* **1,** 1–19, 72–93, 154–157.

VAN HEURCK, H. (1896). "A Treatise on the Diatomaceae", 558 pp., 35 plates, transl. W. E. Baxter. Wm. Wesley & Son, London.

10 | The Study of Lower Tertiary Calcareous Nannoplankton from the North Atlantic Ocean by Means of Scanning Electron Microscopy

A. T. S. RAMSAY

School of Environmental Science, University of East Anglia, Norwich, England

Abstract: Scanning electron microscopy is an ideal technique for the examination of fossil calcareous nannoplankton. The wide field of view, improved depth of focus and increased range of magnification (relative to the light microscope) which are characteristic of the scanning electron microscope allow better qualitative and quantitative assessments of fossil nannofloras than can be achieved by standard optical techniques. Direct viewing of the original sample avoids the selectivity inherent in carbon replication for transmission electron microscopy; while the "three dimensional" image of the electron micrographs and the possibility of viewing the specimen from different angles provides morphological information which is not readily available with the transmission electron microscope.

In situ observations of nannoplankton in soft and indurated sediments show that the processes of lithification and recrystallization modify selectively the species composition of a fossil nannoflora.

A study of the distribution of species in the nannofloras of lower Tertiary samples from the North Atlantic basin indicates that, apart from *Ericsonia ovalis* whose abundance in samples of the *Sphenolithus furcatolithoides* and *Isthmolithus recurvus* zones shows a correlation with latitude, the proportions of species of the genera *Discoaster*, *Rhabdosphaera* and the families Braarudosphaeridae, Pontosphaeraceae, Helicosphaeraceae and Zygolithaceae in the sediments is influenced mainly by the solution of calcium carbonate in relation to depth.

The phenomenon of thickening of species of *Discoaster* is re-examined, and a possible relationship to water temperature is suggested.

INTRODUCTION

The purpose of this article is to discuss the application of scanning electron microscopy to the study of the calcareous Nannoplankton. Examples illustrating the usefulness of this technique for the study of the coccolith-bearing algae are drawn from my own observations of Lower Tertiary species from the North

Atlantic basin. Where relevant, information obtained from samples outside this area is also introduced into the discussion.

Details concerning the samples and sampling localities are given in the Appendix, in Fig. 1, and in Table 3.

CALCAREOUS NANNOPLANKTON: DEFINITION AND HISTORY OF RESEARCH

Calcareous nannoplankton is a collective term which incorporates extant and fossil coccolith producing unicellular pelagic algae, with the extinct discoasters. The coccolith-bearing cells of living species occur in the form of calcified cysts or coccospheres. These have been shown to represent non-motile resting stages in the life history of a species, the motile cells of which bear delicate mineralized or unmineralized organic scales (Black 1969; Parke and Adams 1960). Although the discoasters (Plate I, Figs 1–9) are distinguished taxonomically from the coccoliths at various levels within the suprageneric hierarchy there is general agreement among authors that the discoasters like the coccoliths were produced by chrysophitid algae. Black (1968) suggests that discoasters were possibly also produced by cells in an encysted condition.

The delicate motile stages of planktonic species have not been recorded in the fossil record, and are indeed unrepresented in the bottom deposits of modern oceans. Complete coccospheres or more usually their component coccoliths do fossilize, and together with the discoasters (in pre-Quaternary sediments) are an important constituent in the 1 μ–30 μ size fraction of ancient and recent deep sea sediments.

Although the calcareous nannoplankton has been studied since the middle of the 19th century, interest in the group has expanded considerably during the past 20 years. This expansion can be attributed to two causes, firstly the application of transmission electron microscopy to the systematics of the group, and secondly the demonstration of the value of nannofossils as biostratigraphical indicators in deep-sea and shallow marine sediments by Bramlette and Riedel (1954).

One of the earliest photographs of a fossil coccolith to be taken with the scanning electron microscope was reproduced by Maurice Black in an article in *Endeavour* (1965). Subsequently only six published works based on the observation of fossil calcareous nannoplankton by scanning electron microscopy have appeared. These are concerned with the application of the stereoscan to the study of nannoplankton (Honjo *et al.*, 1967), the taxonomy of Cretaceous and Lower Tertiary calcareous nannofossils (Noël, 1969a,b; Bilal ul Haq, 1968, 1969) and the *in situ* observation of nannoplankton in consolidated and unconsolidated sediments (Noël, 1969c).

With all due respect to these authors and their respective contributions, it would be premature at this stage to suggest that scanning electron microscopy has led to a significant advancement in our knowledge of any aspect of the calcareous nannoplankton. Most of our data concerning the stratigraphy and taxonomy of these organisms is derived from analyses made either with the light or transmission electron microscope. While the most recent attempt at establishing a "natural" classification related to phylogeny (Black, 1968) is based almost entirely on results obtained by transmission electron microscopy (TEM).

THE APPLICATION OF SCANNING ELECTRON MICROSCOPY TO THE STUDY OF CALCAREOUS NANNOPLANKTON

The techniques of scanning electron microscopy (SEM) together with the principles of the instrument have already been discussed by Oatley et al. (1965) and by Oatley (1966). Papers by Hay and Sandberg (1967), Honjo and Berggren (1967), and Honjo and Okada (1968) consider the advantages of this technique in micropalaeontological research. I do not propose to reiterate what these authors have already expressed, and shall confine my comments to the technique as applied to the study of nannoplankton.

Scanning electron microscopy is in many ways better suited for the examination of nannoplankton than either light microscopy or transmission electron microscopy. The preparation of samples for examination in the microscope is as simple as for light microscopy, and merely involves mounting and drying a droplet of suspended particles in the $1~\mu\text{--}40~\mu$ size range on the specimen stub and coating with a conductive metal. The higher range of magnifications and greater depth of focus which are characteristic of the scanning electron microscope, however, allow better qualitative and quantitative assessments of fossil nannofloras than can be achieved by standard optical techniques since the smaller species are more readily identifiable and can be taken into account.

The possibility of viewing the sample directly eliminates the necessity of carbon replication which is essential for the viewing of nannoplankton in the transmission electron microscope (for details of this technique see Black and Barnes, 1961; McIntyre et al., 1967). Direct viewing of the sample avoids the selectivity inherent in this technique, whereby large specimens of high relief may fall off the carbon film leaving no trace other than a large hole. The degree of selectivity in carbon replication is however to some extent a function of the skill of the operator.

Owing to the wider field of view of the stereoscan at lower magnifications (relative to the TEM) scanning electron microscopy is more suited to the quantitative and qualitative analysis of nannoplankton than transmission electron microscopy. Although the maximum resolution achieved by the SEM (200 Å) does not permit observation of ultra-fine surface features, it is quite adequate both for the identification and taxonomic investigation of even the smallest coccolith species. The "three dimensional" appearance of scanning electron micrographs, and the possibility of viewing specimens from different angles provides morphological information which is not so easily obtainable with the transmission electron microscope.

Scanning electron microscopy is undoubtedly a more powerful technique for the quantitative analysis of nannoplankton samples since it eliminates the bias which is introduced by sample preparation for transmission electron microscopy or the low range of magnification of light microscopy. In my experience the optimum magnification for species counts falls in the 2 K range. During a traverse across a specimen stub all species recorded on the visual display unit at any one time are counted.*

In order to determine whether or not species counts within the same subsample or between subsamples of the same sample population varied significantly, three counts of 300 individuals, two within the same subsample, and one of a separate subsample from the same nannoplankton suspension were subjected to analysis by two non-parametric statistical tests. In the first instance the Wilcoxon Matched-Pairs Signed Ranks-Test (Siegel, 1956) was applied to test the null hypothesis that each sample was drawn from the same population among the three groups. In each instance the null hypothesis could not be rejected at $\alpha = 0\cdot 5$ for a two tailed test. As an additional test of significance the Friedman Two-Way Analysis of Variance by Ranks Test was used to test the same null hypothesis among the three samples simultaneously. Again the null hypothesis could not be rejected at $\alpha = 0\cdot 05$. From these results it is concluded that differences observed between species counts do not constitute convincing evidence of a significant difference either in sampling or counting.

1. Taxonomic Investigation

The small size of individual coccoliths (1 μ–15 μ) lies near the limit of optical resolution, and it is only by the application of electron microscopic techniques that sufficient morphological data have become available to interpret phylogenetic series within the coccolith bearing algae. Naturally problems have arisen in comparing forms described by light microscopy and those identified by

---

*At 1 K the large number of specimens recorded on the visual display unit leads to confusion in species counts particularly when it is necessary to resort to higher magnifications for their identification. The 5 K and higher ranges of magnification tend to bias species counts in favour of the smaller species.

electron microscopy or vice versa but they are not insurmountable, and can be resolved by the kind of method outlined by Perch-Nielsen (1967) which allows the simultaneous study of forms using both techniques.

Scanning electron microscopy will not supersede transmission electron microscopy in the study of calcareous nannoplankton, but the increasing application of this technique will, in the future, undoubtedly make a significant contribution to our understanding of the coccolith-bearing algae. Certainly the possibility of observing all species in fossil floras will enhance our knowledge of the groups as a whole, while the three dimensional image and the ability to view the same specimen at different angles provides morphological information which is not readily available to the transmission electron microscope.

The species of *Discoaster* illustrated on Plate I afford an example of the value of the "three dimensional image" for providing information concerning the variation in thickness which occurs within these species. This information is not so obviously conveyed in "two dimensional" transmission electron micrographs, and may be further obscured by the tendency of carbon replicas from specimens with a high relief to flatten. The interpretation of variation in this character is of course subject to personal opinion. Bramlette and Sullivan (1961) distinguish the more delicate species *Discoaster elegans* from the very similar but thicker *Discoaster barbadiensis*; Bystricka (1968) differentiates the more massive subspecies *Discoaster multiradiatus robustus* from the flatter less thickened *D. multiradiatus* in the Palaeogene of the western Carpathians. In my opinion variations in thickness are an example of intraspecific variation, which is ecologically controlled and possibly related to water temperature.

2. Stratigraphical Investigation

The value of the calcareous nannoplankton for the biostratigraphy of the Lower Tertiary was first demonstrated by Bramlette and Riedel (1954). Since that time several authors have contributed to our knowledge of this part of the geologic sequence, but with the exception of Hay and Mohler (1967) and Hay et al. (1967) who have used a combination of transmission electron microscopy and light microscopy, most stratigraphical interpretation is based on studies with the light microscope.

Scanning electron microscopy is well suited for the qualitative analyses of nannofloras, which are used in stratigraphical interpretation, firstly because the sample preparation is rapid and simple, and also because it allows the scanning of literally millions of specimens within the course of a few hours. The latter characteristic is very important since some index species are extremely rare. In the samples examined by the author the index species *Marthasterites tribrachiatus*

Plate I*

Figs 1–3. *Discoaster multiradiatus* Bramlette and Riedel.
Fig. 1. An unthickened form from the *D. multiradiatus* zone of the Lodo formation (sample 21) of California; No. 70.133.14.
×2700.
Fig. 2. A thickened form from core MP 33C, Mid-Pacific; No. 70.133.4.
×3050.
Fig. 3. A thickened specimen from the *D. multiradiatus* zone of Palmer Ridge, N. Atlantic, sample 5981.10; No. 70.133.21.
×3100.
Figs 4–5. *Discoaster barbadiensis* Tan Sin Hok.
Fig. 4. An unthickened form from the *Marthasterites tribrachiatus* zone of the Lodo formation, sample 69; No. 70.129.14.
×1900.
Fig. 5. A thickened specimen from the *Discoaster lodoensis* zone, core Dog 20; No. 69.70.4.
×4300.
Figs 6–7. *Discoaster lodoensis* Bramlette and Riedel.
Fig. 6. An unthickened specimen from the Lodo formation, sample 69; No. 70.129.16.
×2100.
Fig. 7. A thickened form from the *Discoaster sublodoensis* zone, core RC9–55; No. 70.132.5.
×2250.
Figs 8–9. *Discoaster sublodoensis* Bramlette and Sullivan.
Fig. 8. An unthickened specimen from the *D. sublodoensis* zone of Palmer Ridge, sample D. 5608.2 buff sediment; No. 69.64.15.
×4350.
Fig. 9. A thickened specimen from the same zone in core RC9–55; No. 70.132.22.
×3300.

*All Numbers assigned to the specimens are from the collection of negatives of the School of Environmental Sciences, U.E.A.

10. Study of Lower Tertiary Calcareous Nannoplankton

Plate II

Fig. 1. *Discoaster ornatus* Stradner. *D. ornatus* zone of Palmer Ridge, sample D 5968.2; No. 69.58.5.
   × 3400.
Figs 2–3. *Marthasterites tribrachiatus* (Bramlette and Riedel).
Fig. 2. A relatively uncorroded specimen from the *M. tribrachiatus* zone of the Lodo formation, sample 69; No. 70.129.11.
   × 2600.
Fig. 3. A corroded specimen of the same zone from dredge station Gosnold 49–2150; No. 70.129.23.
   × 2100.
Fig. 4. *Marthasterites contortus* Stradner. *M. contortus* zone of Palmer Ridge, sample D.5981.9; No. 69.60.25.
   × 3000.
Fig. 5. *Isthmolithus recurvus* Deflandre. *I. recurvus* zone of core A167–21; No. 70.134.2.
   × 5400.
Figs 6–7. *Sphenolithus furcatolithoides* Locker.
Fig. 6. A relatively uncorroded specimen from the *S. furcatolithoides* zone of core A 150–1; No. 69.72.10.
   × 2500.
Fig. 7. A highly corroded specimen from the same zone at station D 4799, sample D 4799.47, No. 69.84.5.
   × 4700.
Figs 8–9. *Broinsonia cribella* (Bramlette and Sullivan), comb. nov. (= *Coccolithites cribellum* Bramlette and Sullivan). *M. tribrachiatus* zone of the Lodo formation, sample 69.
Fig. 8. No. 70.129.11.
   × 3700.
Fig. 9. No. 70.129.12.
   × 7000, proximal views.

10. *Study of Lower Tertiary Calcareous Nannoplankton*   187

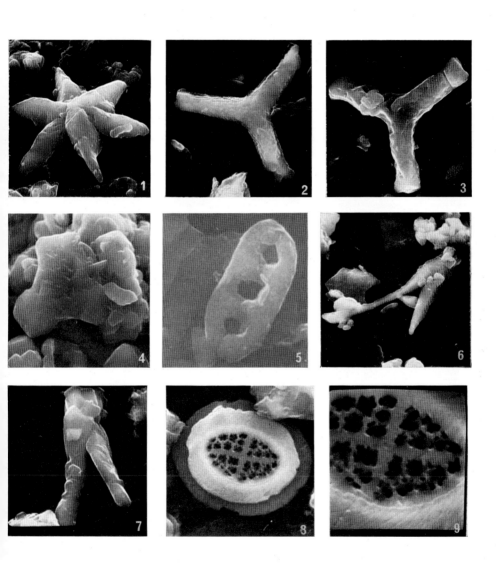

(Plate II, Figs 2–3) and *Isthmolithus recurvus* (Plate II, Fig. 5) are either unrepresented in counts of 300 specimens, or when they are recorded do not exceed 0·6% of the counts. Even at the lowest levels of occurrence most species can be detected by scanning electron microscopy, an added advantage of the technique however is that species which are too small to be identified with the light microscope can also be taken into consideration.

Table I illustrates the calcareous nannoplankton zones identified in the North Atlantic and their correlation with the planktonic foraminiferal zones of Bolli (1957, 1966). The relationship of the North Atlantic samples to the sequence of nannofloral zones is shown in Table 3. (A detailed account of the biostratigraphy of Lower Tertiary North Atlantic sediments is being prepared and will be presented elsewhere.)

Apart from the recognition of three new zones, the *Discoaster ornatus* (Plate II, Fig. 1), *Sphenolithus furcatolithoides* (Plate II, Figs 6–7) and the *Discoaster barbadiensis–Isthmolithus recurvus* partial-range-zone, the sequence of zones presented here is essentially similar to the zones proposed by Hay and Hay and Mohler (in Hay *et al.*, 1967). The *Discoaster ornatus* zone is defined from the last occurrence of *Marthasterites contortus* (Plate II, Fig. 4) to the first occurrence of *Marthasterites tribrachiatus*; the *Sphenolithus furcatolithoides* zone is defined from the first occurrence of the index species to the first occurrence of *Isthmolithus recurvus* (Plate II, Fig. 5). The *Discoaster barbadiensis–Isthmolithus recurvus* zone is defined by that part of the range of *Discoaster barbadiensis* which is concurrent with the earliest part of the range of *Isthmolithus recurvus*.

3. In situ *Investigations*

The majority of investigations devoted to the study of calcareous nannoplankton have been concerned with free specimens in suspensions of the 1 μ–30 μ size fraction of the sediment. *In situ* observations of nannoplankton in hard limestones by transmission electron microscopy have been attempted by several authors (see Flügel and Franz, 1967 for a summary of research up to 1967). Noël (1969c) gives an account of similar observations with the SEM. Once again the sediments require far less preparation than is necessary for transmission electron microscopy, and freshly fractured surfaces can be viewed directly (Plate IV, Figs 5–9).

Although such investigations will undoubtedly provide stratigraphical and palaeontological information from samples which are too hard to disaggregate, an equally important aspect of such investigations is the interpretation of the role of these organisms in the fine fraction of the sediments, and in the influence of post depositional changes in the sediment on the nannoflora.

Table 1. The calcareous nannoplankton zones in the North Atlantic and their relation to the sequence of planktonic foraminiferal zones. * = zones not recorded in the sediments from the North Atlantic.

AGE		ZONATION BASED ON PLANKTONIC FORAMINIFERA After Bolli 1957, 1966 with some taxonomic modification	CALCAREOUS NANNOPLANKTON ZONES
	UPPER	GLOBOROTALIA CERROAZULENSIS	ISTHMOLITHUS RECURVUS
		GLOBIGERAPSIS MEXICANA	ISTHMOLITHUS RECURVUS DISCOASTER BARBADIENSIS
	MIDDLE	TRUNCOROTALOIDES ROHRI *	SPHENOLITHUS FURCATOLITHOIDES
		ORBULINOIDES BECKMANNI	
		GLOBOROTALIA LEHNERI	
		GLOBIGERAPSIS KUGLERI	CHIPHRAGMALITHUS QUADRATUS
		HANTKENINA ARAGONENSIS	DISCOASTER SUBLODOENSIS
			DISCOASTER LODOENSIS
			MARTHASTERITES TRIBRACHIATUS
	LOWER	GLOBOROTALIA PALMERI *	DISCOASTER ORNATUS
		GLOBOROTALIA ARAGONENSIS *	
		GLOBOROTALIA FORMOSA FORMOSA *	
			MARTHASTERITES CONTORTUS
		GLOBOROTALIA REX	DISCOASTER MULTIRADIATUS

Processes which occur during or following sedimentation undoubtedly lead to modifications both in morphology of individuals and in the composition of the flora as a whole. The corroded nature of *Sphenolithus furcatolithoides* and *Marthasterites tribrachiatus* (Plate II, Figs 2–3, 6–7) are examples of a modification which occurred either by the solution of calcium carbonate during deposition, or after burial. Post-depositional modifications are recorded in the species *Ericsonia ovalis* and *Ericsonia alternans* (Plate III, Figs 2–5). In the Lodo formation and shallower water N. Atlantic sediments the complex arrangements of plates on the proximal shields of these species (Plate III, Figs 3 and 5) is obscured by a secondary layer of calcite. This phenomenon is not recorded for the representatives of these species in deep sea sediments. The same process of secondary calcite deposition results in the infilling of the central grid of species like *Broinsonia cribella* (Bramlette and Sullivan) (Plates II and III, Figs 8, 9 and 1). Another characteristic of shallow water sediments is the occurrence of complete coccospheres (Plate III, Fig. 6). These are unrecorded from samples which were collected below the level of the continental shelf, and their occurrence in shallower water sediments may be related to a higher rate of sedimentation.

Lithification and recrystallization are both post-depositional processes which can modify selectively the species composition of a nannoflora. In the highly indurated upper Oligocene Pievepelagos' Marls of the Modena's Apennines (Plate IV, Fig. 5), *Ericsonia ovalis* is the only species preserved. Representatives of this species are, however, in various states of preservation, and range from badly corroded to well preserved coccoliths of which the illustrated form is an example. At station D 5951 the dredge appears to have sampled an intercalated series of thinly bedded soft and hard bands of limestone. In the soft sediments (Plate IV, Fig. 7) the matrix is composed largely of coccoliths, while the $> 60\ \mu$ size fraction is made up of the tests of planktonic foraminifera, radiolarians or fragments of these. In the harder bands (Plate IV, Figs 8–9) the matrix is composed of interlocking minute calcite rhombs, together with the more robust elements of the nannoflora recorded from the softer sediments, i.e. *Ericsonia ovalis* and species of *Discoaster*. The calcite rhombs are similar in size to the coccoliths illustrated on Plate IV, Fig. 7, and it seems likely that they represent recrystallized elements of the nannoflora.

4. Ecological Investigation

The ecology of fossil calcareous nannoplankton has been rather neglected, though several authors have alluded to features of their distribution or morphology which they consider a function of the influence or lack of influence of ecological factors. The following assumptions concerning this group of fossils, however, may be drawn from information published during the past two decades:

(1) That many species have a wide geographical distribution (Bramlette and Riedel, 1954; Bramlette, 1958; Bramlette and Sullivan, 1961; Hay and Mohler, 1967), although their distribution in deep sea sediments is influenced by the solution of calcium carbonate with depth.

(2) Representatives of the families Braarudosphaeridae, Pontosphaeraceae, and of the genus *Rhabdosphaera* Haeckel are rare in deep sea sediments (Martini, 1965).

(3) Species of the genus *Discoaster* Tan Sin Hok are sometimes thickened due to excessive deposition of calcium carbonate (Bramlette and Riedel, 1954; Bramlette and Sullivan, 1961; Martini, 1965).

Judging by reports on calcareous nannofloras from various parts of the globe the first assumption would appear to be reasonably well founded for the Lower Tertiary. Bramlette and Sullivan (1961) consider that this reflects uniformity in the seas of this period. McIntyre and Be (1967) have defined four coccolithophorid floral zones in the recent Atlantic which are a function of the planetary temperature gradient. Although, on the basis of oxygen isotopic analyses of Tertiary benthonic and planktonic foraminifera (Emiliani, 1961), there are reasons for believing that the earth's temperature gradient may not have been as marked during the Lower Tertiary as it is at present, comparisons between nannofloras from areas separated by 10° of latitude or more are probably unreliable if based on a qualitative analysis of the floras. Most of the species considered by McIntyre and Be (1967) have a wide temperature tolerance, and range from tropical to temperate latitudes. Thus variations would not be readily observed from localities within these latitudes on a presence or absence of species basis.

Although the samples included in the present study were obtained from widely separated localities (Fig. 1) the wide range of time they represent (Tables 1 and 3) makes a comparison of total floras between all localities meaningless. Also the selectivity of calcium carbonate solution, discussed in a later paragraph, has influenced the distribution, or, proportion of various species in the sediments thereby obscuring their original relationship to the total nannoflora, and it is difficult to obtain any really meaningful results. One species *Ericsonia ovalis* (Plate III, Figs 4–5) however, whose proportions in the nannofloras (apart from sample V 10–96) is apparently unaffected by calcium carbonate solution shows a positive correlation with palaeolatitude (see Fig. 2) in the *Sphenolithus furcatolithoides* and *Isthmolithus recurvus* zones (see Figs 3 and 4). On the basis of the evidence presented on these graphs it appears that this species

Plate III

Fig. 1. *Broinsonia cribella* (Bramlette and Sullivan). A specimen modified by the secondary deposition of calcium carbonate on the central area, *D. lodoensis* zone, core Dog 20; No. 69.70.3.
× 4800, proximal view.

Figs 2–3. *Ericsonia alternans* Black.

Fig. 2. An unmodified example from the *S. furcatolithoides* zone at core station V18–RD–37; No. 69.44.29.
× 2800, proximal view.

Fig. 3. A specimen modified by the secondary deposition of calcite on the proximal shield, *D. lodoensis* zone, core Dog 20; No. 69.78.55.
× 2500, proximal view.

Figs 4–5. *Ericsonia ovalis* Black.

Fig. 4. An unmodified example from the *Chiphragmalithus quadratus* zone at Palmer Ridge, sample D 5969.7; No. 69.61.25.
× 2550, proximal view.

Fig. 5. A specimen modified by the secondary deposition of calcite on the proximal shield, *D. lodoensis* zone, core Dog 20: No. 69.70.50.
× 3520, proximal view.

Fig. 6. A complete coccosphere from the *Isthmolithus recurvus* zone at core station A 167–21; No. 69.106.64.
× 3000.

Figs 7–8. Representatives of the family Helicosphaeraceae Black.

Fig. 7. *Helicosphaera seminulum* Bramlette and Sullivan, *D. lodoensis* zone, core Dog. 20; No. 69.78.27.
× 8200, proximal view.

Fig. 8. *Lophodolithus nascens* Bramlette and Sullivan, core Dog 20; No. 69.46.42.
× 4800, distal view.

Fig. 9. *Neococcolithes dubius* (Deflandre) a representative of the family Zygolithaceae Noël from core Dog 20; No. 69.48.31.
× 6300, plan view.

10. Study of Lower Tertiary Calcareous Nannoplankton 193

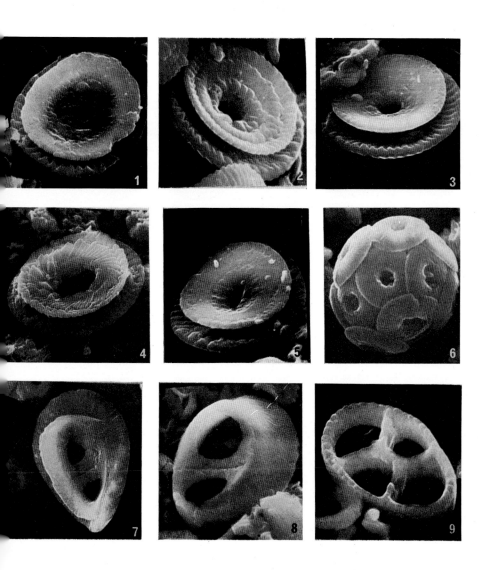

Plate IV

Fig. 1. *Pontosphaera pulcher* (Deflandre), a representative of the family Pontosphaeraceae, *D. lodoensis* zone, core Dog 20; No. 69.70.29.
  × 5600, proximal view.

Figs 2–3. Representatives of the family Braarudosphaeridae Deflandre.

Fig. 2. *Braarudosphaera bigelowi* (Gran and Braarud) Deflandre, *D. sublodoensis* zone of the Lodo formation, sample OC–2; No. 69.96.26.
  × 4300, complete coccolith.

Fig. 3. *Pemma papillatum* Martini. A complete coccolith from the *I. recurvus* zone at core station A 167–21; No. 69.106.51.
  × 2500.

Fig. 4. *Rhabdosphaera perlonga* (Deflandre). A representative of the genus *Rhabdosphaera* Haeckel from the *D. lodoensis* zone at core station Dog. 20; No. 69.78.8.
  × 3800.

Fig. 5. An *in situ* distal view of a well-preserved specimen of *Ericsonia ovalis* Black from the Pievepelagos Marls, Upper Oligocene; No. 70.135.9.
  × 6100.

Fig. 6. A fractured surface of pelagic ooze from the *D. multiradiatus* zone at Palmer Ridge, sample D 5981.4; No. 70.136.29.
  × 1330.

Figs 7–9. Fractured surfaces of pelagic ooze from the *S. furcatolithoides* zone at dredge station D.5951.

Fig. 7. Sample 5951.127, a view of a portion of the sediment showing the matrix of coccoliths and part of a radiolarian; No. 70.136.4.
  × 1240.

Figs 8 and 9. Sample D 5951.101, recrystallized pelagic ooze showing minute rhombs of calcite (? recrystallized coccoliths) and discoasters (Fig. 8), No. 70.136.37.
  × 1500.

Fig. 9. Shows a highly corroded *Ericsonia* sp, and a vein of sparry calcite, No. 70.136.32.
  × 750.

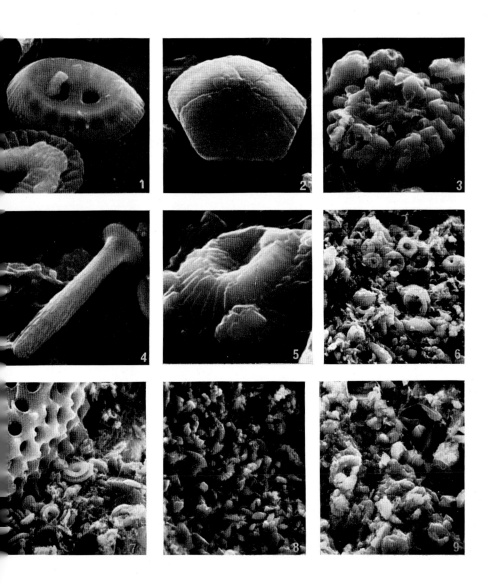

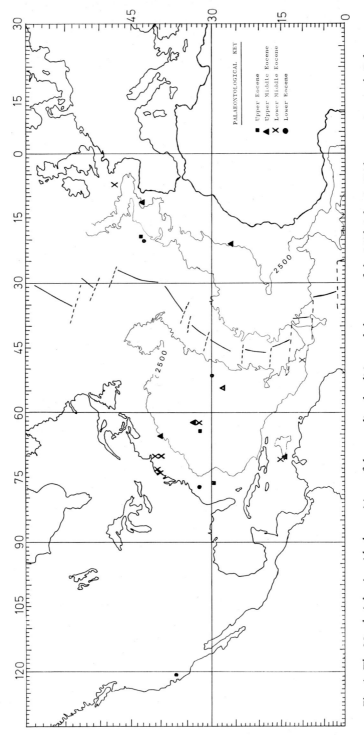

Fig. 1. The North Atlantic with the positions of the sampling localities, and the crest of the Mid-Atlantic Ridge superimposed. Palaeontological ages refer to the oldest fossil occurrence, where samples of more than one age were recovered from a locality. The contoured interval is at 2500 fm (4572 m).

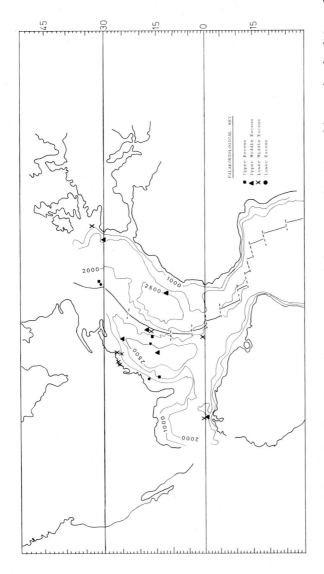

Fig. 2. Reconstruction of the North Atlantic basin of 50 m.y. B.P. with the interpreted positions of the sampling localities, and the crest, of the Mid-Ocean ridge superimposed. The interpretation of the palaeobathymetry is based entirely on morphological criteria, and the inferred positions of the 1000 fm (1828 m), 2000 fm (3657 m) and 2500 fm (4572 m) contours are illustrated.

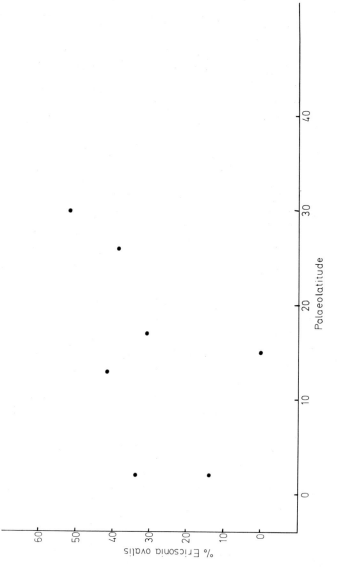

Fig. 3. The relationship between *Ericsonia ovalis* and palaeolatitude in the *Sphenolithus furcatolithoides* zone.

10. Study of Lower Tertiary Calcareous Nannoplankton

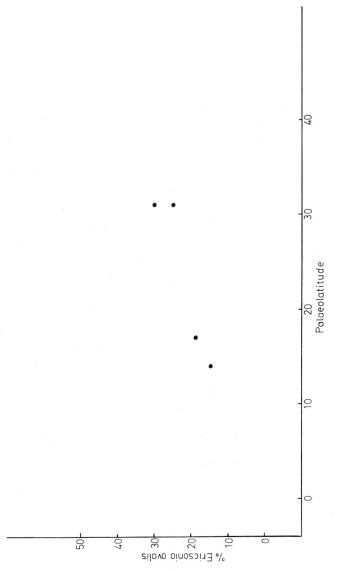

Fig. 4. The relationship between *Ericsonia ovalis* and palaeolatitude in the *Isthmolithus recurvus* zone.

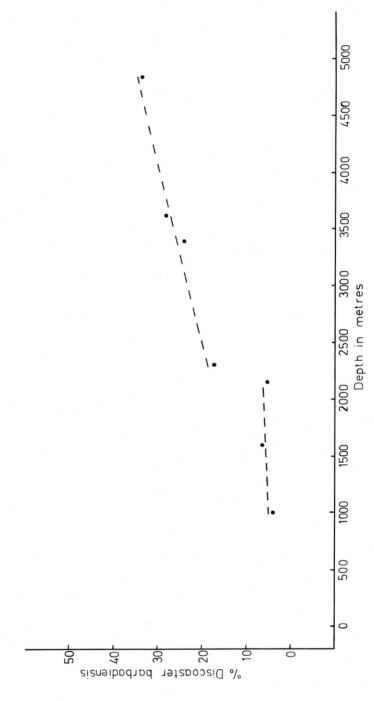

Fig. 5. The relationship between *Discoaster barbadiensis* and sampling depth in sediments of the *Sphenolithus furcatolithoides* zone.

was more abundant in the nannofloras of the more northerly latitudes. Undoubtedly many more examples of latitudinal control will be discovered from quantitative analyses of the more complete floras preserved in the shallower water sediments represented on the continents.

In the recent oceans it has been demonstrated that apart from the uppermost few hundred meters the waters are undersaturated with respect to calcium carbonate at all depths (Peterson, 1966). Berger (1967, 1968, 1970) has shown that the solution of planktonic foraminifera increases with depth; that the process is selective and favours the preservation of the more robust thick shelled species.

In the samples studied the proportions of the more robust species of the genus *Discoaster* vary considerably between localities assigned to the same zone. In order to determine whether the observed variation is depth defined, all species other than *Discoaster barbadiensis* (Plate I, Figs 4–5) were eliminated for analysis in order to exclude differences in abundance which are related to the factor of time.

Figure 5 shows that in the *Sphenolithus furcatolithoides* zone the proportion of *D. barbadiensis* in the samples shows a positive correlation with recent depth. Figure 6 shows a similar correlation between *D. barbadiensis* and depth in all samples from which this species was recorded. It is apparent from this graph that the percentage of this species in the encircled group of samples does not conform to the depths from which the samples were collected. A probable explanation for this discrepancy is that the position of these samples no longer bears any relationship to the position or depth at which they were deposited. In order to test this hypothesis a morphological reconstruction of the lower Tertiary North Atlantic at 50 m.y. B.P. (see Fig. 2) was attempted based on information presented by Funnell and Smith (1968) on the rate of sea floor spreading; the positions of the lower Tertiary latitudes were obtained from Irving (1964). One assumption has been made in the physiographic reconstruction namely that the width of the mid-ocean ridge was of a similar order of magnitude as at present during the lower Tertiary. The interpreted positions of the samples on the reconstructed ocean floor are illustrated on Fig. 2. Although this does not provide information concerning the absolute depth of the samples it does give an indication of the relative depth at which the samples were deposited. The ranked relative depths of the samples is given on Table 2; samples A 16–21 and A 164–25 have, however, been ranked according to their species content which corresponds with samples having a ranked depth of 1 or 2.

The relationship between the proportion of *D. barbadiensis* in the sediments and ranked depth is shown on Fig. 7. The fact that the distribution of this species shows a better correlation with ranked "palaeodepth" than with the sampled

Table 2. Dredge and core localities, depth sampled and ranked depth.

Sample	Present depth (m)	Ranked depth
G 13	13	1
G 28	9	1
RC 9–5	543	2
A 2–5	1006–1262	2
Gosnold 49–2150	1675	2
A 167–21	1455	2
A 164–25	1570	2
Dog 20	1000	3
D 5951	1013–1040	3
D 5981	5282–4542	4
C 10–1	1480	4
A 150–1	1597	4
D 5608	4409	4
D 5623	3869	4
D 5968	5329–4604	4
D 5969	3758–3200	4
V 16–209	4763	5
V 18–RD 37	2166–2496	5
RC 8–2	4625	6
D 4799	2217–2356	6
RC 9–55	3609	7
RC 9–56	3404	7
RC 9–58	3458	7
RC 9–59	3689	7
V 10–96	4839	8

depth (see Fig. 6) supports the hypothesis that depths recorded at the sample stations, in some cases bear little or no relationship to the depth of deposition. The marked increase in the percentage of *D. barbadiensis* in floras between ranked depth 5 and 6 represents a marked increase in the solution of calcium carbonate between these depths which selectively removed the more delicate species, thereby increasing the proportion of the more robust discoasters in the nannoplankton. The approximate upper limit of this depth corresponds with the sampled depth of V 18–RD 37 (2166 m).

The assumption that representatives of the families Braarudosphaeridae, Pontosphaeraceae and the genus *Rhabdosphaera* are restricted to shallower ocean sediments is evident even on the basis of a qualitative analysis of the samples. Members of the family Braarudosphaeridae are, apart from sample A 164–25, restricted to sediments around the continental margin which have a ranked depth of 3 or less.

10. Study of Lower Tertiary Calcareous Nannoplankton 203

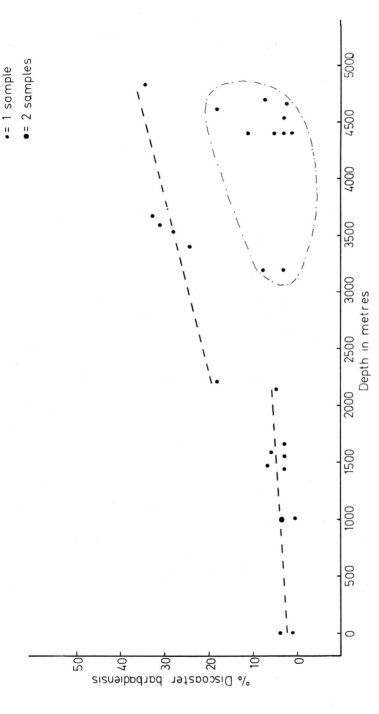

Fig. 6. The relationship between *Discoaster barbadiensis* and sampling depth in all samples from which this species is recorded.

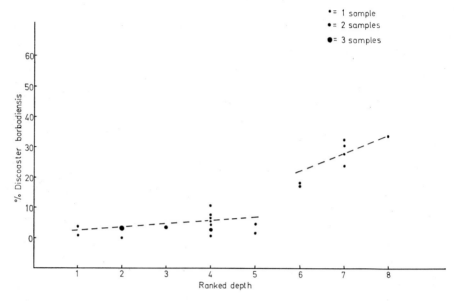

Fig. 7. The relationship between *Discoaster barbadiensis* and relative depths, of the sample as interpreted from Fig. 2. The relative depths of the samples are, apart from samples A 167–21 and A 164–25, ranked according to their positions on the floor of the reconstructed North Atlantic.

Figure 8 shows the relationship between the combined percentages of species of the Braarudosphaeridae, *Rhabdosphaera* and the related families Pontosphaeraceae, Helicosphaeraceae and Zygolithaceae, and depth. Although the proportions of species within an individual member of these groups is to a certain extent related to the age of the sample, the graph does indicate that these groups are an insignificant component (1% or less) of the nannoflora below a ranked depth of 3. Species of the Braarudosphaeridae are absent in sediments below this depth; below a ranked depth of 4 only members of the Helicosphaeraceae are recorded. These results indicate that the proportion of these groups in the sediments is, like the proportion of *D. barbadiensis*, a product of calcium carbonate solution, rather than selectivity on the part of the organisms. In terms of recent depth, solution within these groups increased markedly between 1000 and 1480 m (stations Dog 20 and C 10–1).

The thickening of species of *Discoaster* is evident in most of the Atlantic samples examined, and occurs in the species *Discoaster multiradiatus, D. barbadiensis, D. lodoensis* and *D. sublodoensis* (Plate I, Figs 1–9). In each case the excessive growth of calcite is regular and is clearly a process which occurred

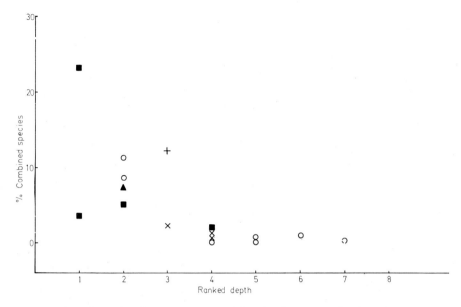

Fig. 8. The relationship between species of the families Braarudosphaeridue, Pontosphaeraceae, Zygolithaceae, Helicosphaeraceae, and the genus *Rhabdosphaera*. The symbols indicate the number of groups recorded at a locality and are ○ = a single group, × = a combination of 2 groups, ■ = 3 groups, + = 4 groups and ▲ = a combination of all groups.

during the life of the species. The same species in samples from the Lodo formation in California do not show evidence of thickening (Plate I, Figs 1, 4 and 6); thus it is likely to have been induced by a factor or factors in the environment. Bramlette and Sullivan record thickening in specimens of the Lodo formation, in samples not examined by the author, which they attribute to some unknown factor in the surface waters which was favourable to the deposition of calcium carbonate. Martini (1965) suggests that thickening only occurred under favourable conditions in equatorial latitudes.

An alternative explanation for this phenomenon is that it is related to water temperature and is an attempt by the cell to adjust its buoyancy to the viscosity of the water. McIntyre and Be (1967) record more heavily calcified representatives of the recent species *Coccolithus huxleyi* from cold water, and it is possible that thickening of the discoasters is an analogous situation. In the *Discoaster multiradiatus* zone the most thickened specimens of *D. multiradiatus* are certainly encountered in the northern latitudes, but until more samples become available from continuous sections and widely spaced localities the problem of thickening must remain an open question.

Table 3. The percentage abundance, based on counts of 300 specimens, of selected species from the sediments of the North Atlantic.

CALCAREOUS NANNOPLANKTON ZONE	DISCOASTER		MULTIRADIATUS	MARTHASTERITES CONTORTUS		DISCOASTER ORNATUS		MARTHASTERITES TRIBRACHIATUS					DISCOASTER LODOENSIS			DISCOASTER SUBLODOENSIS			CHIPHRAGMALITHUS QUADRATUS	SPHENOLITHUS FURCATOLITHOIDES							ISTHMOLITHUS RECURVUS	DISCOASTER BARBADIENSIS	BARBADIENSIS	ISTHMOLITHUS RECURVUS
SAMPLE	V16-209	RC9-5	D5981.10	D5981.9	D5608.4	D5968.2	C10-1	GOSNOLD 49-2150	G 13	D5608.6	D5608.8	AZ-5	Dog 20	G 28	RC8-2	RC9-59	RC9-55	D5608.2	D5969.7	RC9-58	RC9-56	D4799-47	V10-96	A150-1	V18-RD.37	D5951.127	A167-21	A164-25	D5969.2	D5623.3
Discoaster barbadiensis TAN SIN HOK	2.0			3.0	11.0	7.0	7.3	3.0	1.3	3.0	1.3	0.3	4.0	4.0	18.0	3.0	27	13.5	5.3	8.0	2.8	22.4	3.16	7.33	7.6	0.4	7.3	3.7	3.3	3.3
Braarudosphaera bigelowi (GRAN & BRAARUD) DEFLANDRE	1.0											0.6															1.7	1.7		
Micrantholithus attenuatus BRAMLETTE & SULLIVAN		0.3										0.3															1.0			
Pemma papillatum MARTINI																												0.3		
Pontosphaera fimbriata (BRAMLETTE & SULLIVAN)		0.3						1.0					0.3	1.0													7.0	10.0		
Pontosphaera ocellata (BRAMLETTE & SULLIVAN)												0.3																		
Pontosphaera pectinata (BRAMLETTE & SULLIVAN)															10.0															
Pontosphaera plana (BRAMLETTE & SULLIVAN)										2.3			1.3	2.3																
Pontosphaera pulcher (DEFLANDRE)								2.3	0.3			3.3	5.3	7.3				0.3												
Pontosphaera rimosa (BRAMLETTE & SULLIVAN)								0.3	0.3																					
Pontosphaera sp.									0.6						1.2															
Zygodiscus sigmoides (BRAMLETTE & SULLIVAN)						0.6		0.3	0.3	0.3	0.3																			
Zygodiscus sp.									0.3					2.3																
Helicosphaera seminulum (BRAMLETTE & SULLIVAN)	0.3				1.0			0.3		0.3	0.3	0.3																		
Helicosphaera sp.								1.3						0.6	0.3											0.3			0.3	
Lophodolithus nascens (BRAMLETTE & SULLIVAN)															0.6											0.6			0.6	
Neococcolithes dubius (DEFLANDRE)												0.3		0.3																
Rhabdosphaera crebra (DEFLANDRE)														0.6																
Rhabdosphaera perlonga (DEFLANDRE)	0.3											0.6		0.3																
Rhabdosphaera truncata (BRAMLETTE & SULLIVAN)														2.3																0.3

ACKNOWLEDGEMENTS

I am indebted to Dr. Maurice Black, Department of Geology, Cambridge, for many stimulating discussions on various aspects of the calcareous nannoplankton. I also wish to acknowledge Dr. M. N. Bramlette of Scripps Institution of Oceanography for providing material from the Lodo Formation, and Dr. Marta Marucci, Instituto di Geologia, Florence, who provided samples from the Pievepelagos Marls. The research was supported by a N.E.R.C. Research Grant, and was undertaken at the School of Environmental Sciences, University of East Anglia, Norwich.

REFERENCES

BERGER, W. H. (1967). Foraminiferal Ooze: Solution at depths. *Science, N.Y.* **156**, 383–385.
BERGER, W. H. (1968). Planktonic Foraminifera: selective solution and paleoclimatic interpretation. *Deep Sea Res.* **15**, 31–43.
BERGER, W. H. (1970). Planktonic Foraminifera: Selective solution and the lysocline. *Marine Geology* **8**, 111–138.
BILAL UL HAQ, U. Z. (1968). Studies on upper Eocene calcarious nannoplankton from N.W. Germany. *Stockh. Contr. Geol.* **18**, 13–34, Plates 1–11.
BILAL UL HAQ, U. Z. (1969). The structure of Eocene coccoliths and discoasters from a Tertiary deep-sea core in the central Pacific. *Stockh. Contr. Geol.* **21**, 1–19, Plates I–V.
BLACK, M. (1965). Coccoliths. *Endeavour* **24**, 131–137, Plates 1–3.
BLACK, M. (1968). Taxonomic problems in the study of coccoliths. *Palaeontology* **11**, 793–813, Plates 143–154.
BLACK, M. and BARNES, B. (1961). Coccoliths and discoasters from the floor of the South Atlantic Ocean. *J. r. microsc. Soc.* **80**, 137–147, Plates 19–26.
BOLLI, H. M. (1957). Planktonic Foraminifera from the Eocene Navet and San Fernando formations of Trinidad, B.W.I. *Bull. U.S. natn. Mus.* **215**, 155–172, Plates 35–39.
BOLLI, H. M. (1966). Zonation of Cretaceous to Pliocene marine sediments based on planktonic foraminifera. *Boln. inf. Asoc. Venezolana geol. Miner. Petrol.* **9**, 3–32.
BRAMLETTE, M. N. (1958). Significance of coccolithophorids in calcium carbonate deposition. *Bull. geol. Soc. Am.* **69**, 121–126.
BRAMLETTE, M. N. and RIEDEL, W. R. (1954). Stratigraphic value of discoasters and some other microfossils related to coccolithophores. *J. Paleont.* **28**, 385–403, Plates 38–39.
BRAMLETTE, M. N. and SULLIVAN, F. R. (1961). Coccolithophorids and related nannoplankton of the early Tertiary in California. *Micropaleontology* **7**, 129–188, Plates 1–14.
BYSTRICKA, H. (1968). Les Discoastéridés Du Paléogène des Karpates Occidentales. *Acta Geologica et Geographica Universitatis Commenianae, Geologica* **17**, Slovenske Pedagogicke Nakladatelstvo Bratislava 175–243, Plates 59–64.
EMILIANI, C. (1961). The temperature decrease of surface sea-water in high latitudes and of abyssal-hadal water in open oceanic basins during the past 75 million years. *Deep Sea Res.* **8**, 144–147.
FLÜGEL, E. and FRANZ, H. F. (1967). Über die lithogenetische Bedeutung von Coccolithen in Malmkalken des Flachwasserbereiches. *Eclog. geol. Helv.* **60**, 1–17, Plates 1–4.
FUNNELL, B. M. and SMITH, A. G. (1968). Opening of the Atlantic Ocean. *Nature, Lond.* **219**, 1328–1333.

HAY, W. W. and MOHLER, H. P. (1967). Calcareous nannoplankton from early Tertiary rocks at Pont Labau, France, and Paleocene- early Eocene correlations. *J. Paleont.* **41**, 1505–1541, Plates 196–206.

HAY, W. W., MOHLER, H. P., ROTH, P. H., SCHMIDT, R. R. and BOUDREAUX, J. E. (1967). Calcareous nannoplankton zonation of the Cenozoic of the Gulf Coast and Caribbean–Antillean area and transoceanic correlation. *Trans. Gulf-Cst. Ass. geol. Socs.* **17**, 428–480, Plates 1–13.

HAY, W. W. and SANDBERG, P. A. (1967). The scanning electron microscope, a major break-through for micropaleontology. *Micropaleontology* **13**, 407–418, Plates 1–2.

HONJO, S. and BERGGREN, W. A. (1967). Scanning electron microscope studies of planktonic foraminifera. *Micropalaeontology* **13**, 393–406, Plates 1–4.

HONJO, S. and OKADA, H. (1968). Scanning electron microscopy of planktonic Foraminifera: A preparation technique. *J. Fac. Sci. Hokkaido University, Ser.* 4 *Geology and Mineralogy* **14**, 71–76, Plates 17–19.

HONJO, S., NACHIO, M. and OKADA, H. (1967). Study of nannofossils by the scanning electron microscope. *J. Fac. Sci. Hokkaido University Ser.* 4, *Geology and Mineralogy* **13**, 427–31, Plates 51–52.

IRVING, E. (1964). "Palaeomagnetism and its application to geological and geophysical problems." 399 pp. John Wiley and Sons Inc., New York.

MARTINI, E. (1965). Mid-Tertiary calcareous nannoplankton from Pacific deep-sea cores. *In* Submarine Geology and Geophysics (Whittard, W. F., and Bradshaw, R. B., eds.), pp. 393–410. Butterworths, London.

MCINTYRE, A. and BE, A. W. H. (1967). Modern Coccolithophoridae of the Atlantic Ocean—1. Placoliths and cyrtoliths. *Deep Sea Res.* **14**, 561–597, Plates 1–12.

MCINTYRE, A., BE, A. W. H. and PREIKSTAS, R. (1967). Coccoliths and the Pliocene–Pleistocene Boundary. "Progress in Oceanography", Vol. 4: "The Quaternary History of the Ocean Basins", pp. 3–24. Pergamon Press, Oxford.

NOËL, D. (1969a). *Arkhangellskiella* (Coccolithes Crétacés) et formes affines du Bassin de Paris. *Rev. Micropaleont.* **11**, 191–204, Plates 1–3.

NOËL, D. (1969b). Structure de Quelques Coccolithes du Crétacé Supérieur du bassin de Paris examinés au microscope électronique. *Proc. 1st Int. Conf. planktonic microfossils* **2**, 475–485, Plates 1–3.

NOËL, D. (1969c). Etude des coccolithes *in situ* dans les Roches: La Notion de Nannofacies. *Proc. 1st. Int. Conf. Planktonic Microfossils* **2**, 486–491, Plates 1–IV.

OATLEY, C. W. (1966). The scanning electron microscope. *Sci. Prog. Oxf.* **54**, 483–495.

OATLEY, C. W., NIXON, W. C. and PEACE, R. F. W. (1965). Scanning Electron Microscopy. *Adv. Electronics Electron Phys.* **21**, 181–247.

PARKE, M. and ADAMS, I. (1960). The motile (*Crystallolithus hyalinus* Gaarder and Markali) and non-motile phases in the life history of *Coccolithus pelagicus* (Wallich) Schiller. *J. mar. biol. Ass. U.K.* **39**, 263–274.

PERCH-NIELSEN, K. (1967). Nannofossilien aus dem Eozän von Dänemark. *Eclog. geol. Helv.* **60**, 19–32, Plates 1–7.

PETERSON, M. N. A. (1966). Calcite: Rates of dissolution in a vertical profile in the Central Pacific. *Science, N.Y.* **154**, 1542–1544.

SIEGEL, S. (1956). "Nonparametric Statistics for the Behavioral Sciences." 312 pp. Kogakusha Company Ltd., Tokyo.

APPENDIX

Localities of the samples mentioned in the text figures and explanation of plates.

R.R.S. *Discovery*

D 5981: dredge, Palmer Ridge, 42° 51.5′ N, 20° 16.5′ W; 5282–4542 m.
D 5968: dredge, Palmer Ridge, 42° 50.3′ N, 20° 12.2′ W; 5329–4604 m.
D 5608: dredge, Palmer Ridge, 42° 52.0′ N, 20° 16.5′ W; 4409 m.
D 5969: dredge, Palmer Ridge, 42° 55.0′ N, 20° 11.5′ W; 3758–3200 m.
D 5623: dredge, Palmer Ridge, 43° 07.5′ N, 19° 39.5′ W; 3869 m.
D 5951: dredge, 42° 35.8′ N, 11° 57.5′ W; 1013–1040 m.
D 4799: dredge, 26° 0.45′ N, 22° 22.0′ W; 2217–2356 m.

R.V. *Sarsia*

AZ: dredge, 47° 37.0′ N, 07° 31′ W; 1006–1262 m.

Lamont Geological Observatory

G 13: core sample, specimens from bottom at 222 cm, 40° 14.59′ N, 73° 58.58′ W; 13 m.
G 28: core sample, specimens from 44 cm, 40° 14.31′ N, 73° 59.12′ W; 9 m.
Dog 20: core sample, specimens from bottom of core, 39° 43′ N, 70° 43′ W; 1000 m.
V 18–RD 37: dredge, Picket Peak, 39° 40′ N, 65° 55′ W; 2496–2166 m.
A 150–1: core sample, specimens from bottom at 148 cm, 33° 42′ N, 62° 30′ W; 1597 m.
C 10–1: core sample, specimens from bottom at 750 cm, 33° 40′ N, 62° 30′ W; 1480 m.
RC 9–5: core sample, specimens from bottom at 139 cm, 32° 19′ N, 77° 39′ W; 543 m.
A 164–25: core sample, specimens from 360 cm, 32° 13′ N, 64° 31′ W; 1570 m.
V 16–209: core sample, specimens from bottom at 540 cm, 30° 0′ N, 51° 52′ W; 4763 m.
A 167–21: core sample, specimens from bottom at 365 cm, 29° 49′ N, 76° 35′ W; 1455 m.
V 10–96: core sample, specimens from bottom at 735 cm, 27° 52′ N, 54° 38′ W; 4839 m.
RC 9–55: core sample, specimens from bottom at 1026 cm, 14° 42′ N, 70° 53′ W; 3609 m.
RC 9–56: core sample, specimens from bottom at 703 cm, 14° 38′ N, 70° 52′ W; 3404 m.
RC 9–59: core sample, specimens from 762 cm, 14° 37′ N, 70° 50′ W; 3698 m.
RC 9–58: core sample, specimens from bottom at 460 cm, 14° 33′ N, 70° 49′ W; 3458 m.
RC 8–2: core sample, specimens from bottom at 750 cm, 11° 12′ N, 48° 05′ W; 4625 m.

U.S. Geological Survey

Gosnold 49–2150: dredge, 39° 46′ N, 69° 44′ W; 1675 m.

Scripps Institution of Oceanography

MP 33C: core sample, near top centre of Hess Guyot, 17° 49′ N, 174° 17′ W; 1673 m.

Lodo Canyon California

Lodo formation sample 69
Lodo formation sample OC–2
The sample numbers are those used by Bramlette and Sullivan (1961).

Pievepelago (Modena's Appenines) Italy

Pievepelagos Marl (upper part of the Macigno Sandstone).

11 | Fine Details of Some Fossil and Recent Conifer Leaf Cuticles

M. C. BOULTER
Department of Biological Science, North East London Polytechnic, London, England

Abstract: It has been realized for some time that the features of the cuticle of conifer leaves can be used to discriminate between genera with much the same effect as is achieved by using reproductive structures. Since so many fossil conifers are preserved only as vegetative remains, the leaf cuticle is of especial palaeobotanical importance.

With the light microscope a well preserved fossil cuticle may show a good deal of detail of both the structure and arrangement of epidermal cells and stomata. SEM studies, however, have revealed additional structural details which have not been easily discernable by other techniques. Some examples of these fine details are: the surface structure of the cuticle (e.g. the nature of verrucae, papillae, pegs, striations), the nature of buttresses and cavities in the cuticular flanges along the epidermal cell walls, the cuticular flanges between the guard cells and the subsidiary cells, the shape of the lignified thickening of the guard cell walls, and the nature of the stomatal depression within the hypodermis.

When used in conjunction with established light microscope techniques, the scanning electron microscope (SEM) presents new opportunities not only to increase the number and the clarity of definition of cuticle characters available for taxonomic separation, but also to increase our understanding of conifer evolution.

THE USE OF FOSSIL CUTICLES

There are two factors which make the plant cuticle eminently suitable as a subject of study for the palaeobotanist. Firstly, the cuticle has a very stable chemical structure which not only provides the living plant with resistant outer protection, but is also able to survive many biological and geological processes through time. This inertness is due to the chemical stability of the constituent hydroxy monocarboxylic acids such as phloionic acid ($C_{18}H_{34}O_{16}$), cutinic acid ($C_{13}H_{22}O_3$) and cutic acid ($C_{26}H_{50}O_6$). By carboxyl-hydroxyl linkages these are thought to produce polymeric esters which are likely to have ester linkages to lignin, cellulose and tannins, which are thought to be common constituents of plant cuticles (Robinson, 1963). Secondly, the cuticle reflects the surface structure of the epidermal layer which it covers, so presenting details of the epidermal cells, stomata, papillae, glands and other epidermal features. Light microscope study of cuticles commonly shows differences in thickness rather plainly (whether the cuticle is its natural brown or is stained) but only when it is rather thick can one study it as a three-dimensional layer, perhaps with outwardly projecting papillae and inwardly projecting flanges along the cell margins. These projections have often caused palaeobotanists to complain when studying thick cuticles because they give a confused picture. With a thin cuticle optical microscopy shows only paler or darker parts, but though careful studies have been made, examination has been difficult and the conclusions have not always seemed fully convincing to the author making them. With the SEM one has separate and perfectly clear views of the surface projections on the two sides. Features which were resolved doubtfully by careful focusing using an oil immersion objective immediately become unequivocal and obvious. When used in conjunction with results obtained by the light microscope from using macerated cuticles and thin sections, the new information can give valuable comparisons of the change in three-dimensional structure of the leaf epidermis through geological time.

Using the light microscope, figures are usually given at magnifications of a few hundred times only. Magnifications of $\times 1000$ are unusual, presumably because authors consider that the appearances observed at such high magnifications are too vague or not clearly significant. With the SEM however, magnifications of $\times 2000$ or higher are easy to obtain and show features perfectly clearly. It does of course remain to be proved which of these fine features are of taxonomic value, but very many look as though they might be.

In the past, leaf cuticles from fossil plants have been used not only to identify genera and species but also to show some of the variation in epidermal structure that occurs in closely related groups of plants. Recently, in work on the European Tertiary for instance, Rüffle (1963) and Litke (1965) have described fossil plant assemblages consisting of angiosperm leaf remains which they were able to identify by the use of cuticular characters. These features have been used previously in the identification of fossil angiosperm leaf remains in the German browncoal; this work was carried out by Weyland and published as a long series of papers (see, for example, Weyland, 1957). One of the most severe handicaps that beset all these workers was the inadequate source of reference material of Recent angiosperms. For smaller groups such as the cycads however, the situation is otherwise; exhaustive studies on modern material (Pant and Nautiyal, 1963; Greguss, 1968) are available for comparison with fossil forms, though these are most commonly found in the Mesozoic and so the cuticles are not directly similar to those from modern genera (Harris, 1961, 1964). The leaf cuticles from plants of some other gymnosperm groups that are represented by Recent species, namely the conifers, taxads and a few other isolated genera, have been described in detail by Florin for both Recent (Florin, 1931) and some fossil species (Florin, 1951, 1958).

In all these cases, information obtained from the study of fossil leaf cuticles can be successfully employed for the purposes of identification. The reliability of the cuticle as a source of taxonomic characterization has been shown by Stace (1966) for the angiosperms, and by Florin (1958) for the conifers. The fact that this anatomically obscure plant part shows consistent variation between taxa, so as to confirm the character differences of plant groups that are displayed by their reproductive organs, demands further comment.

Many experiments have been performed on living plants to show that environmental changes can effect the morphology of cuticular characters. For example, Watson (1942) demonstrated that the degree of waviness on the anticlinal walls of epidermal cells is dependent upon the incident light intensity. Allsopp (1954) showed that both the abundance and arrangement of stomata, as well as the size and form of the ordinary epidermal cells, were effected by variation in the concentration of nutrient glucose. Though in experiments such as these there are rather gross changes of life conditions leading to considerable variation in the epidermis, it would appear from Florin's work on a comparison of *Taxus* from many countries (Florin, 1958), that ordinary environmental differences have rather slight effects on the cuticles. Thus while environmental causes cannot be discounted, we have no reason as yet to think that they are important. Some differences, such as good growth conditions leading to large leaves and large epidermal cells, can be discounted, since the proportion of stomata to epidermal cells (stomatal index) remains remarkably constant.

METHODS

By varying the severity of the maceration process, more information can be obtained from the leaf cuticles than by the use of the conventional Schulze's solution/alkali method alone. Mild maceration of Recent leaves using concentrated nitric acid only, for 2–3 hours, or Schulze's solution only, for 5–15 minutes, usually leads to the retention of the lignified guard cell thickenings (Boulter, 1970) and, in some genera such as *Pinus*, parts of the hypodermis. The normal use of Schulze's solution followed by 10 per cent ammonium hydroxide solution produces cuticle with no lignified parts attached. There is scope for further study of the effects of maceration techniques. The substance of the lignified fibres in general is destroyed quicker than the lignified parts of the guard cells, whilst the cuticle itself is eventually dissolved by prolonged acid treatment.

Little attention has been given in light microsope studies to the structure of the guard cell wall. This is probably due to its having been removed from the cuticle preparations by alkali treatment, though in a great deal of fossil material the structures are not preserved.

In the same way that no universal technique can be described for the preparation of fossil pollen from rock samples of different types, no single technique for the preparation of fossil cuticles can be regarded as ideal for all cases. The variation in the type of fossilization, with such complications as differential mineral replacement and charring (prior to fossilization), require different kinds of treatment. Of the fossil material studied here, that from the Derbyshire Neogene beds (Boulter and Chaloner, 1969) was soaked either in 40 vol hydrogen peroxide for about one month, or else macerated more quickly with concentrated nitric acid alone, for up to ten minutes. Other material, from the Eocene Bagshot beds (Chandler, 1962), and the Rhenish Hauptflöz of Miocene age (Kilpper, 1968), was subjected only to the latter process. In all this material there seemed to be little difference in the cuticles prepared by either method, though the peroxide technique was more easily controlled due to the greater time involved.

In the present study, the prepared cuticle was attached to the SEM specimen stub with Durofix and coated with gold-palladium. Recently, Heslop-Harrison (1970) has successfully examined the surface of the contemporary angiosperm leaf *Pinguicula grandiflora* without coating. This was found to be an unsuccessful method for dealing with conifer cuticles, perhaps due to their much greater thickness.

CHARACTERS USED IN LIGHT MICROSCOPE INVESTIGATIONS

Information on leaf cuticle structure obtained by SEM studies can be most useful when interpreted in conjunction with conventional examination of both cuticle macerations and thin leaf sections by light microscopy. It is, therefore, appropriate here to summarize briefly some of the more important cuticular characters which have been used by previous workers in both Recent and fossil conifer cuticles.

One of the characters most widely used in studies of both Recent (Florin, 1932) and fossil conifers (e.g. Mädler, 1939; Florin, 1951 and 1958; Szafer, 1961; Chandler, 1962 and 1963) is the distribution of the stomatal apparatus in relation to the ordinary epidermal cells, and whether they are more or less random or arranged in rows. Most commonly, the bifacially flattened lanceolate leaves—

type II of Laubenfels (1953)—are hypostomatic (stomata only on the lower surface) whilst those leaf types that are not flat in section—types I, III and IV of Laubenfels—are usually amphistomatic (stomata present on all surfaces). There is also variation in the orientation of the stomata in relation to the axis of the leaf.

The number of different cell types making up the stomatal apparatus is another important criterion that has been widely used to identify fossil leaf material (Florin, 1958). These features can be best understood if the epidermal cells in the fossil material are preserved in a petrified state for examination in thin section. In conifers, the two guard cells which are often depressed within the leaf surface are surrounded either by a variable number of subsidiary cells in a single row around the guard cells (monocyclic), or by subsidiary cells plus one complete or incomplete row of encircling cells (dicyclic); occasionally there are two such bands of encircling cells (tricyclic). In some genera, either subsidiary cells or encircling cells are shared by the same stomatal apparatus. Whichever of these arrangements is present, a subsidiary cell has the same initial cell as its radially neighbouring encircling cell(s). This haplocheilic development, in which the mother cell of the guard cells has no close ancestry with that of the subsidiary cells, is typical of all conifers (Florin, 1951). The limits within which these characters are developed by a particular genus allow quite accurate generic identification from cuticular material alone.

Other features which have been used in these studies include striate markings on subsidiary cells and epidermal cells; either straight or undulating anticlinal walls; the presence of hairs, teeth, papillae and crystals; and the occasional presence of cuticular protrusions from the outer cuticle surface extending into the outer part of the stomatal depression. These take the form of finger-like projections or lappetts (Townrow, 1965), or else a raised ring or "Florin ring" (Buchholz and Gray, 1948), and occur among living conifers most especially in species of *Podocarpus* and *Pinus* (Fig. 1) which have substantially depressed guard cells. Unfortunately some of these miscellaneous features are unreliable taxonomically since their occurrence is not only affected by the environmental conditions of the plant and by the maturity of the leaf, but their structure can become distorted by the maceration process.

Though no exhaustive survey of the cuticles of Recent conifer species has been attempted, such information as we have (e.g. Florin, 1931) shows that certain species of a genus may differ sufficiently in cuticular characters to be recognized apart with ease. But others are certainly much more similar and we do not know whether they can be separated by the cuticle alone. This limitation has had most influence on the identification of Tertiary conifers where the significance of specific identity (especially in comparison with living species) is

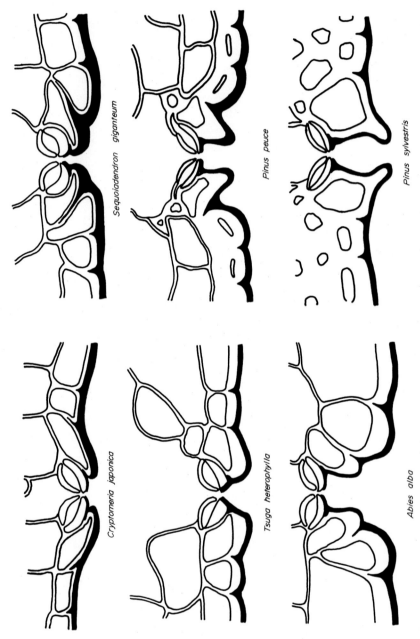

Fig. 1. Drawings of transverse median stomatal sections of six species of modern conifers, based on thin sections examined by light microscope. The extent of the inner face of the cuticle (solid black) was determined by SEM studies.

11. Details of Leaf Cuticles from Fossil Conifers

greater than in other fossil material. But this has not deterred some authors from identifying fossil material in terms of a living species solely from the evidence of macerated cuticles (Szafer, 1961). On the other hand, recent precise application of these light microscope cuticle techniques to fossil material previously thought to have affinities to modern *Abies*, shows that the leaves are best considered as a new genus, *Tritaenia* (Mägdefrau and Rudolf, 1969).

FINE DETAIL CHARACTERS

Most of the conventional characters of the conifer leaf cuticle which are mentioned in the previous section can also be studied by the SEM. In the majority of cases, the higher resolution and depth of focus which the instrument makes available discloses finer details of these known characters, while some completely new features are revealed for the first time. The work has also led to a reinterpretation of the three dimensional structure of some of these previously known features.

Under the light microscope, stomatal arrangement and the number of subsidiary and encircling cells are very prominent features, and because of their constancy they are rated highly as taxonomic characters in comparison to other cuticular features. Although these are easily observed by the SEM, other regu-

---

Table 1. List of fossil and Recent conifer taxa whose macerated leaf cuticles have been examined in this study by SEM. The specimens of *Sequoia couttsiae* were collected by Miss M. E. J. Chandler from the Bagshot beds, Dorset, and are deposited in the British Museum (Nat. Hist.) (V 44503); the leaves of *Cryptomeria rhenana* were collected by Dr K. Kilpper from the German Miocene; while the remaining fossil material was collected by the author from the Neogene plant beds in Derbyshire. All the living material was obtained from the forest plots at Bedgebury arboretum.

Sequoia couttsiae Heer	*Sequoiadendron giganteum* (Lindley) Buchholz
Cryptomeria rhenana Kilpper	*Cryptomeria japonica* Don
Cryptomeria sp.	*Metasequoia glyptostroboides* Hu & Cheng
Abies sp.	*Sciadopitys verticillata* (Thurnberg) Siebold & Zuccarini
Pinus sp.	*Abies alba* Miller
	Picea abies (L.) Karsten
	Pinus sylvestris L.
	Pinus peuce Grisebach
	Tsuga heterophylla (Rafinesque) Sargent

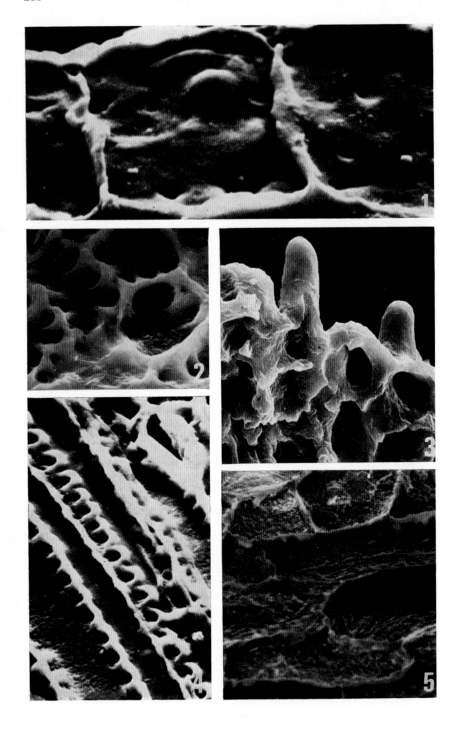

larly occurring cuticular characters are evident which are seen less clearly in the light microscope. As a result, the SEM can change the morphological and taxonomic importance that we give to a particular character. This extension of the range of taxonomically useable features in leaf epidermis studies will greatly enhance the value of the cuticle to both the palaeobotanist and to the systematist.

1. Structure of the Inner Cuticle Surface in Ordinary Epidermal Cells

In general the leaf cuticle of gymnosperms is thicker than that in the leaves of plants from other groups, so that many of the markings which appear on one surface of the cuticle are not found on the other. The inner cuticle surface is attached to the cellulose wall of the epidermal cells; this wall is thought to serve as the medium through which cutin precursors pass on their way from the site of synthesis in the plant to the leaf surface (Lee and Priestley, 1924). In most of the genera studied by SEM (see Table 1), the inner surface is quite smooth, though in all of the taxodiaceous material both Recent and fossil, and in *Tsuga heterophylla*, it has a granular character, due perhaps to the method by which the cutin precursors finally leave the epidermal cell wall microfibrils (Plate II, Figs 2 and

---

Plate I

Fig. 1. Fossil *Abies*. Ordinary epidermal cell cuticle from the upper leaf surface, showing a verruca and several pores.
$\times 2000$.

Fig. 2. *Metasequoia glyptostroboides*. Intercellular cuticular flanges from the upper leaf surface, showing buttresses and cavities which, before maceration, were filled with cellulose.
$\times 1000$.

Fig. 3. *Sciadopitys verticillata*. A broken edge of the lower leaf surface showing papillae from the stomatal band and their basal pegs.
$\times 1500$.

Fig. 4. *Pinus peuce*. Cuticle from the cells bordering a row of stomata (top right) showing intercellular flange buttresses (bottom left) and balustrades in the cuticle of the epidermal cells at the edge of the depressed stomatal ridge (passing from top left to bottom right).
$\times 1000$.

Fig. 5. Fossil *Sequoia couttsiae*. Cuticle from the ordinary epidermal cells, showing striations on the inner surface.
$\times 1500$.

All photographs are by SEM and show the inner surface of macerated leaf cuticles.

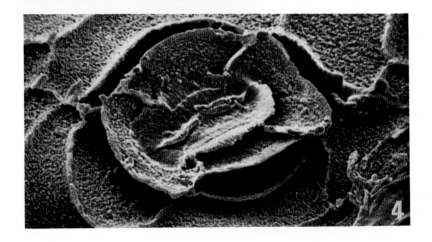

4). However, *Sequoia couttsiae* shows longitudinal striations as well as this granular character (Plate I, Fig. 5). Many fossil leaf cuticles show fine striae in the light microscope (for example, Harris (1969) frequently refers to a fine "mottling" of the epidermal cell cuticle in some of the Yorkshire Bennettitales) but it is not possible to say if it is external or internal or even in the thickness of the cuticle, without the aid of a SEM.

Both in *Abies alba* and in the fossils of *Abies* (see Table 1) the inner surface of the leaf cuticle commonly shows two characters: verrucae about 1 μ in diameter, and pores no greater than 0·1 μ in diameter (Plate I, Fig. 1). When present, there are either one, two or three verrucae present on each cell area, and they are distributed in an apparently random manner. Similar structures have been observed on the outer cuticle surface of *Cordaites* from the German Carboniferous; this work, by Wartmann (1969) is the first published study of fossil cuticles seen by SEM. Most commonly, there are three to six pores per cell area in both species of *Abies* that were studied here (Plate I, Fig. 1); these are most likely to connect the pits in the outer secondary wall of the abutting epidermal cell to the exterior of the leaf. Comparable structures have been described in the epidermis of the leaf of *Allium cepa* from transmission electron microscopy (TEM) studies (Scott *et al.*, 1958), although Juniper (1958) was unable to observe any such features (by carbon replica TEM studies) to explain the source of wax crystals present on the leaf surface of various angiosperms.

SEM investigations of *Metasequoia glyptostroboides* leaf cuticle show that small

---

Plate II

Fig. 1. Fossil *Sequoia couttsiae*. A fully macerated stoma. The intercellular flanges between the subsidiary cells and the guard cells, and the encircling cells and the subsidiary cells are particularly prominent.
×1500.

Fig. 2. Fossil *Cryptomeria rhenana*. A fully macerated stoma. That part of the cuticle surface resembling the inside of an opened umbrella is the cuticular flange separating the subsidiary cells from the encircling cells.
×1000.

Fig. 3. Fossil *Pinus* sp. A row of stomata showing the guard cell walls, and to the right, the cuticle of a row of balustrade epidermal cells.
×500.

Fig. 4. *Sequoiadendron giganteum*. A fully macerated stoma showing similar features to the photograph of *Cryptomeria rhenana* above.
×1000.

All photographs are by SEM and show the inner surface of macerated leaf cuticles.

"papillae" protrude into the cellulosic layer of both the subsidiary cells and the ordinary epidermal cells in the stomatal region (Plate III, Fig. 3). Florin (1958) speaks of these cells as "papillate", but this term is normally used for a solid or hollow outwardly pointing projection. These inwardly directed "papillae" are clearly of a very different nature; whether or not they could have been recognized under the light microscope, they certainly were not in this case.

2. Intercellular Flanges

At the boundary of neighbouring epidermal cells, the cuticle's inner surface intrudes to produce a projecting system of flanges which reflect the distribution of all the outer epidermal cells. Although most knowledge of the macerated cuticles from both Recent and fossil leaves is based on the study of these patterns, examination of the details of the flanges under light microscope has only been rewarded by interpretations of much uncertainty. From the preliminary work that is reported here, it can be seen that the fine structure of the flange itself is very variable within the conifers.

The cuticular flange marking the outline of the ordinary epidermal cells is shown by light microscope to be either "undulating" or straight. The SEM demonstrates many different types of "undulating" wall; the undulations in living conifers are mainly due to buttress structures penetrating the base of the flange, as seen in *Pinus peuce* (Plate I, Fig. 4), and fossil *Abies* (Plate III, Fig. 2). These buttresses are particularly prominent in flanges from the upper leaf surface in *Metasequoia glyptostroboides* where they often form originally cellulose-filled cavities in the flange between neighbouring epidermal cells (Plate I, Fig. 2).

Although dependent upon the thickness of the cuticle itself (which varies with both maturity and environment) both the depth and the thickness of the intercellular flange show a taxonomic consistency. Leaves of *Sequoia couttsiae* have a relatively deep and narrow flange in comparison to the other members of the Taxodiaceae that have been examined (Plate I, Fig. 5 and Plate II, Fig. 1). In the fossil *Abies* the walls near the stomata are represented by a double flange (Plate III, Fig. 2), suggesting cutinization at the sides of the middle lamella but not replacing the substance of the middle lamella itself. As far as I know, no such double cuticular flange has been reported, but it may indeed be widespread. Harris (1964) for example, described the "cell outlines" of *Nilssonia tenuinervis* as consisting of "a thin strip flanked by thicker borders", and of *N. tenuicaulis* as appearing "double as though thin along the middle". These may well represent similar double flanges.

Florin (1958) has described unusually elongate epidermal cell outlines on the cuticles from some *Pinus* species. These have also been seen in these SEM investigations, although at higher resolution thin poorly developed flanges are observed separating the supposedly elongate cells into units of similar area to the neighbouring equilateral cells.

3. The Cuticle in the Region of the Stomatal Apparatus

This is widely regarded as one of the most informative parts of the leaf cuticle for both morphological and taxonomic purposes. It is particularly important for examination by SEM that the cuticle is fully macerated; the hypodermis, particularly in genera such as *Pinus* where it is several cells thick (Fig. 1), must be completely removed, by prolonged acid treatment if necessary.

The outer surface of the cuticle in this region frequently reveals a simple undifferentiated stomatal aperture under which the underlying guard cells are situated, though some genera show interesting specialization. Several Mesozoic conifers have papillae projecting from the outer surface of the subsidiary cells almost to the centre of the stomatal depression. These can be recognized by light microscope work (e.g. Lorch, 1968; Mägdefrau and Rudolf, 1969) as well as by SEM investigations (Alvin, 1970). In the Lower Tertiary plants *Podocarpus setiger* and *Microstrobus sommervillae* from Tasmania (Townrow, 1965), the stomatal pit is overhung by broader more flap-like cuticular extensions, or lappetts. Further, in some fossil types and in some Recent species of *Pinus* the outer surface of subsidiary cells or encircling cells (whichever are at the upper edge of the stomatal depression) is raised to give a ring-like projection at the outer entrance to the stomatal depression. This Florin ring can be seen in thin sections of *Pinus peuce* leaf epidermis (Fig. 1). When viewed in SEM it is seen to be made up of as many lobes as there are subsidiary cells (the subsidiary cells cannot be recognized from the outer cuticle surface where there are no flanges). Each lobe joins with its neighbour to form the complete ring with an undulating surface. The SEM is particularly suitable for the investigation of characters such as these; by studying a variety of fossil types of different ages some meaningful evolutionary sequence for this surface character might be detected; the increase in specialization from papillae to lappetts and then to the Florin ring may have some phylogenetic significance.

The inner surface of the fully macerated cuticle shows many features of the guard cells, subsidiary cells, and encircling cells which are not visible by light microscope techniques. But the examination by light microscope of thin sections, differentially stained with Sudan IV to show cutin, has helped in interpreting the results to be discussed here. There is a substantial variation within the conifers in the depth of the guard cells within the leaf tissue. This affects the appearance of the inner cuticle surface as seen by the SEM. For instance, *Tsuga*

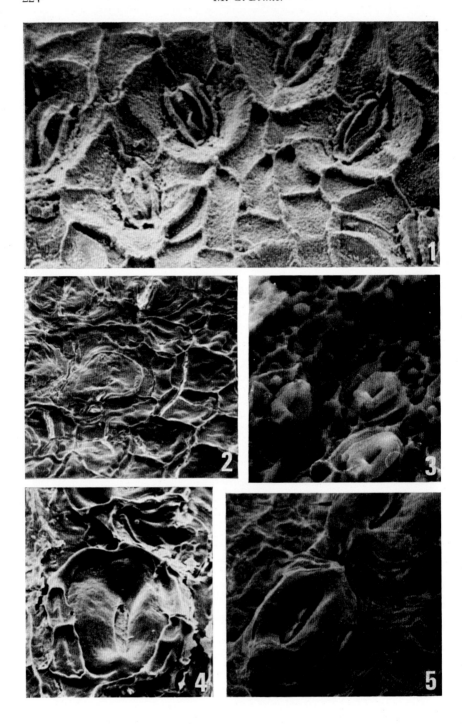

heterophylla has its guard cells just below the surface (Fig. 1); the subsidiary cells of adjacent stomata are often shared and there are seldom any encircling cells. *Sequoiadendron giganteum* (Fig. 1) is usually di-cyclic (Florin, 1958) with the subsidiary cells as well as the guard cells depressed below the leaf surface. In the SEM the macerated cuticle of *Tsuga heterophylla* shows the guard cell cuticle to be only slightly depressed (i.e. raised upwards when viewed from the inner surface), while that of the subsidiary cells is on the same plane as the ordinary epidermal cells (Plate III, Fig. 1; text Fig. 1). In *Sequoiadendron giganteum*, however, the depressed cuticle of the subsidiary cells, rather in the shape of an opened umbrella, bear the guard cell cuticle at their centre, and themselves lie below the cuticle of the encircling cells (Plate II, Fig. 4; Fig. 1). Similar interpretations of the photographs of fossil material show that the subsidiary cells of *Sequoia couttsiae* (Plate II, Fig. 1) were at the leaf surface and the guard cells were only slightly sunken within the leaf (Fig. 2). A different situation can be seen in the fossil *Cryptomeria rhenana* (Plate II, Fig. 2) which has guard cells more sunken than in the modern species (Figs 1 and 2).

The flanges of cuticle between the guard cells and the subsidiary cells, and the subsidiary cells and the encircling cells are much more evident in the SEM than in light microscope sections where they are barely resolved. The depth and thickness of these flanges show consistent differences in many of the genera

---

Plate III

Fig. 1. *Tsuga heterophylla*. Fully macerated stomatal region showing the subsidiary cells to be at the leaf surface.
 $\times$ 600.

Fig. 2. Fossil *Abies* sp. Compressed guard cell walls and ordinary epidermal cells. These latter cells (right) show verrucae and compressed buttresses at the intercellular flanges.
 $\times$ 500.

Fig. 3. *Metasequoia glyptostroboides*. Stomatal region showing lignified guard cell walls and cuticular pegs on both subsidiary cells and ordinary epidermal cells.
 $\times$ 400.

Fig. 4. Fossil *Pinus* sp. Compressed guard cell walls, surrounded by a single row of small lignified (?) cells. The inner wall (projecting uppermost in this photograph) of the two cells on the left has broken off.
 $\times$ 1000.

Fig. 5. Fossil *Pinus* sp. The guard cell walls are not compressed. Between the cuticle and these guard cells lie the lignified cells which are seen more clearly in Fig. 4.
 $\times$ 1000.

All photographs are by SEM and show the inner surface of macerated leaf cuticles.

studied. In *Sequoiadendron giganteum* (Plate II, Fig. 4) for instance, the flange between guard cells and subsidiary cells is much more prominent than in *Cryptomeria rhenana* (Plate II, Fig. 2), even allowing for the compression of the fossil cuticle.

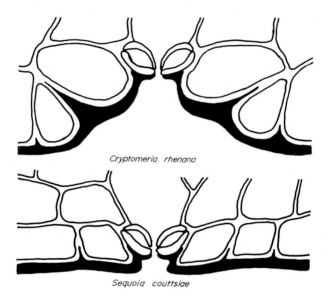

Fig. 2. Reconstructions of transverse median stomatal sections of two fossil types, based on the evidence of SEM studies of macerated cuticles (Plate II, Figs 1 and 2). In neither case are the cell walls of the epidermis and hypodermis preserved for examination in T.S. The guard cells of *Sequoia couttsiae* are only slightly depressed, much as in the modern *Cryptomeria japonica* or *Tsuga heterophylla* (cf. Fig. 1). The guard cells and the subsidiary cells of *Cryptomeria rhenana* however are depressed below the leaf surface, in the same way as those of the modern *Sequoiadendron giganteum* (cf. Fig. 1).

4. The Thickened Guard Cell Wall

When thin sections of conifer leaf material are stained in a lignin stain such as phloroglucin and concentrated hydrochloric acid, the guard cells show a positive reaction. Such lignin thickenings of Recent gymnosperms were studied a good deal by some of the earlier students of fossil cuticles but sometimes with unfortunate results, since the stained epidermis of the living plant (with strongly coloured lignin thickening) was compared with fossil cuticles in which no lignin at all was present. More recently, the lignin thickenings of fossil stomata have occasionally been figured, for example in *Glossopteris* leaves from Tanganyika (Pant, 1958). His preparations represented whole epidermal layers which for some reason were preserved in isolation, not as true cuticles. When he prepared the cuticle, the lignified thickening vanished. Such specimens have been met occasionally in other formations but we have no method of producing them from ordinary fossils. Working with conifer material however, Florin (1931) refers to the thickenings in drawings of his thin sections, and Pant and Srivastava (1968) comment on the effect of maceration on the lignified guard cells in *Araucaria*. The guard cells of cycad leaves also have a lignified thickening (Pant and Nautiyal, 1963; Thomas and Bancroft, 1913), while in the Bennettitales the guard cells have strong cuticular thickenings (Harris, 1932) which look very like the lignin thickenings of other gymnosperms. We cannot say whether these cutin plates were accompanied by lignin.

A leaf of *Sequoiadendron giganteum* was partially macerated so as to retain just a few of the lignified guard cells attached to its cuticle; in the SEM this shows the different stages of their disintegration (Plate IV, Fig. 1). In some stomata, only the outer part of the guard cell wall remains, while in others the entire wall is either present or absent. The appearance of the lignified guard cell thickenings of some members of the Taxodiaceae as seen in the light microscope has been reconciled with the structure's form as seen in the SEM (Boulter, 1970). It has been shown that there is sufficient variation in the morphology of this structure for it to be used as an aid to the identification of fossil *Cryptomeria* leaves. The most important character of the lignified guard cells in the Taxodiaceae is shown to be the structure of the thickening at the joined polar ends of the guard cells, referred to as the polar lamellae. A preliminary SEM examination of these guard cell thickenings in other conifer genera shows that there is a specifically consistent variation. *Abies alba* and a fossil species of the genus from the Derbyshire Neogene beds show a much reduced polar lamella, with prominent separation of the end walls of the guard cells (Plate IV, Fig. 5). Two Recent species of *Pinus*, *P. peuce* (Subgenus *Haploxylon*) and *P. sylvestris* (Subgenus *Diploxylon*) have each guard cell thickening united at the polar ends so there is no external evidence of two separate cells; each pair of polar lamellae is a single united structure tapering to a single more or less pointed end (Plate IV, Fig. 2). Fossil leaves of this genus from Derbyshire show a similar type of polar differentiation (Plate III, Fig. 5), as do the guard cells of *Sciadopitys verticillata*. In all the other genera that have been studied, there is an intruding ridge between each pair of guard cells, quite prominent in *Picea abies* and *Abies alba* but less so in the Taxodiaceae. It is perhaps relevant to compare the absence of a flange between the guard cells in *Pinus* and *Sciadopitys* to the situation that has been shown to exist in some grass guard cells by Brown and Johnson (1962). They used TEM techniques to show that the polar ends of the guard cell walls are

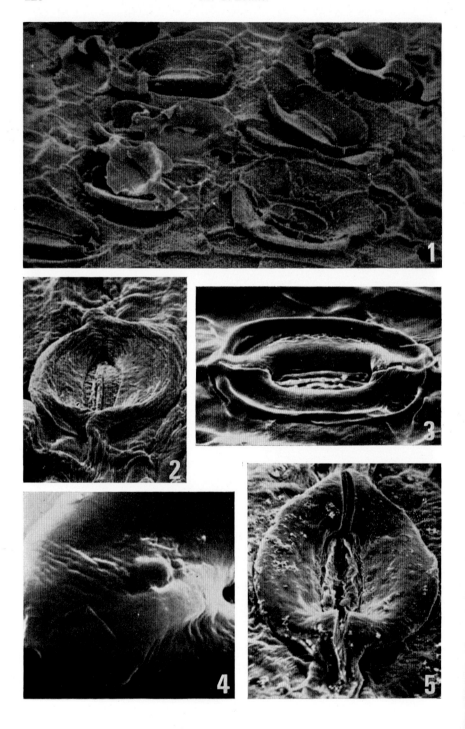

Plate IV

Fig. 1. *Sequoiadendron giganteum*. Partially macerated stomatal region showing the thickened walls of the guard ... either completely removed (bottom right) or both upper and lower parts remaining (top right) or the lower part (uppermost in the photograph) removed. The depressed subsidiary cells have their cuticle, when viewed from the inner surface, above that of the encircling cells (cf. Plate II, Fig. 4).
× 500.

Fig. 2. *Pinus peuce*. Lignified cell wall, with a striate surface ornament.
× 1000.

Fig. 3. *Picea abies*. Lignified guard cell wall with no polar lamellae.
× 1000.

Fig. 4. *Sequoiadendron giganteum*. Region of the lignified guard cells which lies between the polar lamellae (out of the picture to the bottom left) and the stomatal aperture (the edge of this is just visible on the right). It is suggested (page 230) that the folds seen here may play some part in the opening and closing mechanism of the stomatal aperture.
× 5000.

Fig. 5. *Abies alba*. Lignified guard cell wall, showing the polar lamellae (bottom, centre) and the clear separation of the two guard cells.
× 1000.

All photographs are by SEM and show the inner surface of macerated leaf cuticles.

incomplete, so as to give confluent protoplasts. Comparable TEM techniques would be needed to show that this situation exists in conifers too.

Different types of surface ornamentation on the lignified guard cell walls have been recognized in some genera. In most of the species examined the surface is quite smooth with no pores or pits visible, though *Pinus peuce* shows striations on the surface (Plate IV, Fig. 2). Light microscope investigations of some macerated angiosperm cuticles shows comparable surface features (Ziegenspeck, 1955) termed micellae, which it is suggested, may play some part in the mechanism of stomatal movement. High magnification by SEM of the polar region of *Sequoiadendron giganteum* guard cells show a series of folds on the surface of the cells at the curved part between the polar lamellae and the stomatal aperture (Plate IV, Fig. 4). These could mark the region of movement within the guard cell wall that leads to the opening and closing of the stomatal aperture; further observations of living material would be needed to verify this suggestion.

It is often asserted that conifer guard cells, like those of other modern groups of higher plants, are differentially thickened on the inside of the stomatal aperture, to form ridges which ensure the tight closure of the stoma. Furthermore, Schwendener (1881) suggested that especially thin parts of the guard cell walls called Hautgelenke are present on the subsidiary cell side of the guard cell wall, to serve as hinges for stomatal movement. SEM studies of conifer stomata show no evidence of such ridges on the inside of the aperture, just as there is no sign of Hautgelenke on the surface in contact with the subsidiary cells. But there are two regions of greatly reduced thickening passing longitudinally along each guard cell, one on the inner edge, the other on the outer. This is not only seen in light microscope sections (Fig. 1), but also in the differentially macerated cuticle of *S. giganteum* as seen by SEM (Plate IV, Fig. 1). These areas may be equivalent to Hautgelenke, and allow the two thickened parts of the guard cell walls to move upwards and downwards (in respect to the leaf surface) rather than sideways. It might be assumed that the polar lamellae provide the movable parts of the lateral guard cell wall with continual support, their being firmly fixed between neighbouring encircling cells, or subsidiary cells, and the hypodermis.

5. *Specialization of Depressed Stomata*

Thin sections of conifer leaves show that in some genera such as *Pinus*, *Picea* and *Abies* the hypodermis is two or three cells deep and the sunken guard cells are at the hypodermal level (Fig. 1). In *Tsuga heterophylla* and members of the Taxodiaceae, hypodermal fibres are more or less absent from the stomatal region; indeed, they are rarely more than two cells thick elsewhere around the leaf. It is more difficult to examine cuticle macerations of leaves from the first of these types by SEM since the hypodermal fibres are often not completely removed by the maceration procedure, or else the depressed part of the cuticle becomes distorted and collapsed. Nevertheless there are features present in leaf types with a substantial hypodermis in the stomatal region which can be conveniently studied by SEM.

SEM photographs from the undermacerated cuticles of *Picea abies*, *Pinus peuce* and fossil *Pinus* from the Derbyshire Neogene reveal a row of lignified cells attached to the outer edge of the guard cells (Plate III, Fig. 4). This is particularly evident in the fossil material where the cells concerned are not covered by the guard cells, perhaps as the result of compression. These results suggest that the composition of the walls of these cells is nearer to that of the guard cells than to the cells of the hypodermis, if only for the fact that they survive the strong acid maceration in the same way as the guard cells and not the hypodermal fibres.

Another interesting feature of *Pinus* leaf cuticle that is seen only in SEM may be mentioned here. The rows of stomata that are present on the leaves are situated at the bottom of grooves on the leaf surface. The cuticle at the edge of these grooves appears in SEM to have strengthening balustrade-like thickenings extending into the epidermal cells right along the edge of the groove (Plate I, Fig. 4). This feature has been seen in both the modern species of *Pinus* that have been studied, and in the Derbyshire fossil leaves of this genus, where compression has slightly distorted the appearance of the thickenings (Plate II, Fig. 3).

In *Sciadopitys* leaves, the guard cells are not deeply depressed within the leaf tissue, though they are aggregated in a stomatal band and protected by stout papillae up to 45 μ in length, which project from the subsidiary cells and encircling cells. Examination by SEM of these papillae from a broken edge of the inner surface of the cuticle (Plate I, Fig. 3) shows each papilla to have a basal lump of cutin which protruded into the cellulose wall of the subsidiary cell.

FUTURE APPLICATIONS

At the beginning of this century Porsch (1905) wrote that "though there is a general type of stoma characteristic of the gymnosperms as a whole, there is one special type for almost each genus when the finer structure is taken into consideration". With the aid of the SEM even greater detail of the stomata can be seen to facilitate, at least in some genera, identification at the specific level. Preliminary results show that as much detail can be obtained from a well preserved fossil cuticle as from that of Recent material. But if the fossil leaf has been coalified and compressed, then many of the characters suitable for recognition of the species are lost. In the specimens of *Abies* leaf from Derbyshire for instance, details of the guard cell walls are difficult to interpret because they are squashed onto the cuticle of the subsidiary cells (Plate III, Fig. 2).

Despite these difficulties the instrument will be useful for the identification of fossil material, especially when there is either little material available, or when the suspected identification from preliminary light microscope investigations is controversial. Recently for instance, leaf material from the North American Tertiary has been assigned to the Section *Stachycarpus* of the genus *Podocarpus*, solely on the basis of vegetative characters (Dilcher, 1969). Not only is this the first record of that Section of the genus *Podocarpus* from North America, but also few parts of the plant were available for study—two reasons for extreme caution in making this identification. In such circumstances, the additional evidence from SEM studies will give valuable confidence to the palaeobotanist.

By giving separate pictures of the inner and outer surface of the cuticle, the SEM shows various things clearly which were at best very difficult with light microscopy and were not considered at all by many authors. Very probably, some of these features could not have been resolved by light microscopy at all, or only by awkward methods such as delicate focusing under high magnification. Features like deeply projecting cell flanges, particularly when oblique, which make optical microscopy difficult, become easy and are observed in the most direct way possible. Then too, the new instrument's ease in handling magnifications well over a thousand, and its very great depth of focus, give it enormous advantages.

How far the new facts brought out by this superior microscopy will affect conifer taxonomy remains to be seen, but already it has contributed facts which look as though they should be valuable. The doubt which exists about the taxonomic position of many specimens, for example, Dilcher's *Podocarpus*, is due in part only to the novelty of the identification; such an important find should have all the confirmation possible. The SEM should be able to add at least a few facts which might help one way or the other.

The new instrument also has a significant part to play in the detection, back through the fossil record of closely related species, of particular cuticular characters which show slight phenotypic change between species. A comparison of the slight changes that occur in some details of the fine structure of conifer cuticles gives an opportunity for the first time to examine the evolutionary development of subcellular features in a quantitative manner. The Taxodiaceae for instance, not only have a diverse history in the Tertiary and Mesozoic, but they also have leaf cuticles well suited to examination by SEM. Measurements of the range of cell size, the angle of the subsidiary cell cuticle surface to the leaf surface, the depth of the intercellular flange between the guard cells and the subsidiary cells, the size of the polar lamellae on the guard cell wall, and similar features, will make interesting comparisons within this group. Such quantitative analyses may enable palaeobotany to make more important contributions to our understanding of the mechanism of the evolutionary process than it has done in the past.

ACKNOWLEDGEMENTS

I would like to thank the Trustees of the British Museum for allowing me to prepare specimens of their *Sequoia couttsiae* for the SEM, the Directors of Bedgebury Arboretum for conifer material from their forest plots, and Dr K. Kilpper for specimens of *Cryptomeria rhenana*.

Dr M. Muir and Dr K. Alvin very kindly enabled me to use the SEM at Imperial College, and together with Dr J. Townrow and Dr W. G. Chaloner gave much help and enthusiasm to my understanding of fossil cuticles. I especially thank Dr Chaloner and Prof. T. M. Harris for critically reading the manuscript.

REFERENCES

ALLSOPP, A. (1954). Experimental and analytical studies of pteridophytes, 24. Investigations on *Marsilea*, 4. Anatomical effects of changes in sugar concentration. *Ann. Bot., London.* **173**, 1032–1035.

ALVIN, K. L. (1970). Study of fossil leaves by Scanning Electron Microscopy. *3rd Annual Symp. Scanning Electron Microscopy*, pp. 121–128. Illinois Institute of Technology Press, Chicago.

BOULTER, M. C. (1970). Lignified guard cell thickenings in the leaves of some modern and fossil species of Taxodiaceae (Gymnospermae). *Biol. J. Linn. Soc.* **2**, 41–46.

BOULTER, M. C. and CHALONER, W. G. (1970). Neogene fossil plants from Derbyshire, England. *Rev. Palaeobot. Palynol.* **10**, 61–78.

BROWN, W. V. and JOHNSON, C. (1962). The fine structure of the grass guard cell. *Am. J. Bot.* **49**, 110–115.

BUCHHOLZ, J. and GRAY, N. (1948). Taxonomic review of *Podocarpus*, I. *J. Arnold Arbor.* **29**, 49–63.

CHANDLER, M. E. J. (1962). "The Lower Tertiary Floras of Southern England", Vol. II, 176 pp. British Museum (Natural History), London.

CHANDLER, M. E. J. (1963). "The Lower Tertiary Floras of Southern England", Vol. III, 169 pp. British Museum (Natural History), London.

DILCHER, D. L. (1969). *Podocarpus* from the Eocene of North America. *Science, N.Y.* **164**, 299–301.

FLORIN, R. (1931). Untersuchungen zur Stammesgeschichte der Coniferales und Cordaitales. *K. svenska Vetensk Akad Handl.* (Ser. 5), **10**, 1–588.

FLORIN, R. (1951). Evolution in Cordaites and Conifers. *Acta Horti Bergiani* **17**, 257–402.

FLORIN, R. (1958). On Jurassic Taxads and Conifers from north-western Europe and eastern Greenland. *Acta Horti Bergiani*, **17**, 257–402.

GREGUSS, P. (1968). "Xylotomy of the Living Cycads, with a Description of Their Leaves and Epidermis", 260 pp. Akadémiai Kiadó, Budapest.

HARRIS, T. M. (1932). The fossil flora of Scoresby Sound east Greenland, 2: description of seed plants *incertae sedis* together with a discussion of certain cycadophyte cuticles. *Meddr. Grønland.* **85, 3** 1–112.

HARRIS, T. M. (1961). "The Yorkshire Jurassic Flora, I", 212 pp. British Museum (Natural History), London.

HARRIS, T. M. (1964). "The Yorkshire Jurassic Flora, II", 191 pp. British Museum (Natural History), London.

HARRIS, T. M. (1969). "The Yorkshire Jurassic Flora, III", 186 pp. British Museum (Natural History), London.

HESLOP-HARRISON, Y. (1970). Scanning Electron Microscopy of fresh leaves of *Pinguicula*. *Science, N.Y.* **167**, 172–174.

JUNIPER, B. E. (1958). Growth, development, and effect of the environment on the ultrastructure of plant surfaces. *J. Linn. Soc. (Bot.)* **56**, 413–419.

KILPPER, K. (1968). Koniferen aus den Tertiären Deckschichten des Niederrheinischen Hauptflözes, 3. Taxodiaceae und Cupressaceae. *Palaeontographica B*, **124**, 102–111.

LAUBENFELS, D. J. DE (1953). The external morphology of coniferous leaves. *Phytomorph.* **3**, 1–20.

LEE, B. and PRIESTLEY, J. H. (1924). The plant cuticle, 1. Its structure, distribution, and function. *Ann. Bot.* **38**, 525–545.

LITKE, R. (1965). Kutikularanalytische Untersuchungen im Niederlausitzer Unterflöz. *Paläont. Abh. B*, **2, 2** 330–426.

LORCH, J. (1968). Some Jurassic conifers from Israel. *J. Linn. Soc. (Bot.)* **61**, 177–188.

MÄDLER, K. (1939). Die pliozäne Flora von Frankfurt am Main. *Abh. senckenb. naturforsch. Ges.* **446**, 1–202.

MÄGDEFRAU, K. and RUDOLF, H. (1969). Morphologie und Histologie der Nadel der Conifere *Abietites linkii* (Roem.) Dunk. aus dem Wealden den Hils. *N. Jb. Geol. Paläont. Mh.* **5**, 288–298.

PANT, D. D. (1958). The structure of some leaves and fructifications of the *Glossopteris* flora of Tanganyika. *Bull. Br. Mus. (Nat. Hist.) Geol.* **3**, 125–175.

PANT, D. D. and NAUTIYAL, D. D. (1963). Cuticle and epidermis of recent Cycadales. Leaves, sporangia and seeds. *Senck. biol.* **44**, 257–347.

PANT, D. D. and SRIVASTAVA, G. K. (1968). On the cuticular structure of *Araucaria* (*Araucarites*) *cutchensis* (Feistmantel) comb. nov. from the Jabalpur series, India. *J. Linn. Soc. (Bot.)* **61**, 201–206.

PORSCH, O. (1905). "Der Spaltöffnungsapparat im Lichte der Phylogenie." Jena.

ROBINSON, T. (1963). "The Organic Constituents of Higher Plants", 306 pp. Burgess, Minneapolis.

RÜFFLE, L. (1963). Die obermiozäne (sarmatische) Flora vom Randecker Maar. *Paläont. Abh., B*, **1, 3**, 139–296.

SCHWENDENER, S. (1881). Über Bau und Mechanik der Spaltöffnungen. *Sber. Akad. Wiss. Wien* **21**, 833–867.

SCOTT, F. M., HAMNER, K. C., BAKER, E. and BOWLER, E. (1958). Electron microscope studies on the epidermis of *Allium cepa. Am. J. Bot.* **45**, 449–461.

STACE, C. A. (1966). The use of epidermal characters in phylogenetic considerations. *New Phytol.* **65**, 304–318.

Szafer, W. (1961). Miocene flora from Stare Gliwice in Upper Silesia. *Prace Inst. Geol.* **33**, 1–205.

Thomas, H. H. and Bancroft, N. (1913). On the cuticles of some Recent and fossil cycadean fronds. *Trans. Linn. Soc. Lond.*, **8**, 155–204.

Townrow, J. A. (1965). Notes on some Tasmanian pines. 1. Some Lower Tertiary podocarps. *Pap. Proc. R. Soc. Tasm.* **99**, 87–107.

Wartmann, von R. (1969). Studie über die papillen-formigen Verdickungen auf der Kutikule bei *Cordaites* an Material aus dem Westfal C des Saar-Karbons. *Argumenta Palaeobotanica*, **3**, 199–207.

Watson, R. W. (1942). The effect of cuticular hardening on the form of epidermal cells. *New Phytol.* **41**, 223–229.

Weyland, H. (1957). Kritische Untersuchungen zur Kutikularanalyse tertiärer Blätter, III. Monocotylen der rheinischen Braunkohle. *Palaeontographica B*, **103**, 34–74.

Ziegenspeck, H. (1955). Das Vorkommen von Fila in radialer Anordnung in den Schliesszellen. *Protoplasma* **44**, 385–388.

12 | Scanning Electron Microscopy of Fungi and its Bearing on Classification

LILIAN E. HAWKER

Department of Botany, University of Bristol, Bristol, England

Abstract: The value of the scanning electron microscope in the study of structure and development in fungi and as an aid to their classification is discussed.

A number of examples are given of studies of fungus spores and their surface ornamentation and the significance of these in classification is indicated.

The examples chosen represent most of the main taxonomic groups of the fungi and include members of the Mucorales, hypogeous Ascomycetes, Deuteromycetes, Ustilaginales (smut fungi) and hypogeous Gasteromycetes. Examples are also given of the use of the scanning electron microscope (SEM) in the elucidation of the method of development of fungus spores and their mode of attachment to the parent structure.

The desirability of parallel studies with the light microscope and various techniques with the transmission electron microscope as an aid to interpretation of scanning electron micrographs is emphasized.

INTRODUCTION

The scanning electron microscope has been little used in the study of fungi. At the XIth International Botanical Congress held at Seattle, U.S.A. in 1969 only three of the mycological papers read were concerned with the use of this instrument, whereas, a whole symposium and a number of contributed papers presented results obtained by the use of the transmission electron microscope (TEM).

The reason for this apparent neglect of the scanning electron microscope by mycologists is not far to seek. The fragile nature of most fungus structures and the high water content of their vegetative hyphae severely limits the usefulness of any technique for their examination which involves desiccation without prior embedding. Thus the preparation of fungus material for viewing with the scanning electron microscope frequently leads to the shrinkage, distortion or complete collapse of hyphae and other fragile structures. Occasionally, as will be shown later, slight or moderate shrinkage actually yields additional information and thus aids interpretation of particular structures. In general, however, it is a severe disadvantage. Attempts to solve the problem of shrinkage and collapse of delicate hyphae by rapid freeze drying of the material before coating have not in my experience proved successful but further attention to modifications in preparing such material for viewing is desirable.

Many fungus structures are, however, sufficiently rigid to retain their shape during drying and coating. These include many, but by no means all, spores together with sclerotia, rhizomorphs and some particularly hard textured stromata and fruit bodies. Spores have received most attention, not only as a result of their importance in classification, particularly at the generic and species levels, but because it is often difficult to resolve details of their surface ornamentation by light microscopy, particularly when dark pigment is present in the spore wall.

Willetts (1968) used the scanning electron micrope in conjunction with light microscopy in a study of the development of the stromata of the brown rot fungi (*Sclerotinia* spp) of rosaceous fruits and of the outer surfaces of the sclerotia of three unrelated fungi (1969). Such large objects, however, are normally observed satisfactorily with the light microscope alone, the degree of magnification provided by the SEM being unnecessary.

Other possible uses of the SEM which have so far been little exploited, are the examination of soil organisms (including fungi) *in situ* (Gray, 1967), the study of penetration of plant cuticles or of insect integuments by fungal pathogens and the observation of epiphytic organisms on plant surfaces.

STRUCTURE AND DEVELOPMENT OF SPORES

1. Surface Ornamentation of Fungus Spores

Fungus spores are often very beautiful objects and the many kinds of surface ornamentation, spines, warts, ridges, flanges, reticulations etc. can be better seen with the aid of the SEM than by any other available means. The practical advantages of such studies are the increased precision with which details of spore structure and development may be ascertained, thus enhancing their value in classification.

(a) Zygospores of the Mucorales. Zygospores of *Mucor* and related genera are usually described as "warty" with no further details. The SEM shows clearly that the warts are of different shape in different genera and in species of the same genus. This is illustrated in Plate I, Figs 1–3 which shows the regular flat-topped warts of *Rhizopus sexualis* (Smith) Callen and the fluted pyramidal warts of *Zygorhynchus moelleri* Vuillemin. Those of *Mucor* spp, not illustrated, are different again, having fluted warts which are somewhat flattened and wider at the base than those of *Zygorhynchus*. The smoothness of the surface of zygospores of

Absidia and *Phycomyces* is confirmed by the SEM, which with some spores (e.g. the basidiospores of *Hysterangium*, p. 246) has revealed fine ornamentation not visible by light microscopy.

(b) Sporangiospores of *Rhizopus* spp. Ellis and Heseltine (1969) used the scanning electron microscope to study details of the surface configuration of sporangiospores of species of *Rhizopus* and reported clear specific differences. Here the results with this microscope, as with my own results with zygospores, satisfactorily confirmed the existing classification.

(c) Conidia of *Cunninghamella* spp. Young (1968) studied spores of *Cunninghamella* and other members of the Mucorales by the carbon replica technique. In a study of the germination of spores of *Cunninghamella* (Hawker, Thomas and Beckett, 1970) the conidia were examined with the SEM and Young's results with this genus largely confirmed although the SEM gave less distortion than the carbon replica process. An interesting point emerging was the resemblance in form of the spines of *Cunninghamella* conidia to those on the sporangium surface in species of *Mucor*. Both were finely tapering and had a hemispherical swelling at the base (Plate II, Figs 1 and 2). This supports the hypothesis put forward by many mycologists that conidia such as those of *Cunninghamella* have evolved from multispored sporangia by gradual reduction in number of spores to one. The outer wall of such a single-spored sporangium would thus be homologous with the wall the sporangium of a multi-spored species. The similarity in surface ornamentation between the conidia of *Cunninghamella* and the sporangia of *Mucor* is thus significant.

(d) Ascospores of *Elaphomyces granulatus* Fr. The species of *Elaphomyces* produce their fruit-bodies underground and are mycorrhizal in habit. The mature fruit-bodies contain masses of ascospores which in most species are black, dark brown or purple. The characters of these, including surface ornamentation, have been used in species determination. Lange (1956) and Eckblad (1962) proposed that *E. granulatus* should include only specimens with spiny spores and that those with an irregular warty surface should be referred to *Elaphomyces asperulus* Vittad. Reference to the early literature relating to *Elaphomyces* (Vittadini 1831, 1842; Hollós, 1911) suggested that such a division of *E. granulatus* and *E. asperulus* was not entirely satisfactory.

The writer had long been interested in hypogeous fungi and had in her possession a large collection of specimens from the British Isles, Norway and N. America. These were re-examined together with specimens from the herbaria of Berkeley, Broome (Berkeley and Broome 1837–1885, 1846) and others at the British Museum (Natural History) and the Royal Botanic Gardens, Kew. It became clear that specimens which were otherwise identical had different types of spore ornamentation and that specimens with spores showing similar surface features were not always similar in other characters.

From notes made at the time of collection of the writer's material it seemed likely that the age of the specimens, the soil conditions at the time of collection and the rate at which they were subsequently dried could all influence the final form of the ornamentation (Hawker et al., 1967). The use of the scanning microscope enabled the method of development of the spores to be examined further and confirmed the fact that specific differences in this genus could not be based on spore ornamentation alone. Light microscopy and transmission electron microscopy (Fig. 1.) had shown that the young spores of a number of species of *Elaphomyces* were at first smooth in outline, hyaline and thin-walled. They were shed from the ascus in an immature condition and immediately afterwards began to enlarge, acquire thickened walls and later pigmentation. The thick walls while still hyaline, could be seen to consist of a number of radial rods embedded in a matrix. With the development of pigmentation the fate of these rods could not be determined by light microscopy alone, and, owing to the hard, brittle nature of the impermeable thickened spore walls, fixing, embedding and sectioning for observation with the TEM was unsatisfactory. The SEM was, however, ideal for studying such hard black spores and showed that the final form of the spore surface depended on whether the radial rods seen in young spores remained separate, forming spines, or became clumped together to form irregular warts (Plate II, Figs 3, 4 and 5). It was concluded that the degree of clumping of the rods could be influenced by rate of drying of the matrix in which the rods were originally embedded. This was confirmed experimentally. As a result it was suggested that *E. granulatus* Fr. and *E. asperulus* Vitt. should be divided on peridial character rather than on spore ornamentation.

Plate I

Scanning electron micrographs

Fig. 1. Mature Zygospore of *Rhizopus sexualis*, showing ornamentation with warts shaped like irregular truncated cones or pyramids. Fragments of exospore remain adhering to these warts. One of the two suspensors is seen at the top of the picture.

Fig. 2. Mature Zygospore of *Zygorhynchus moelleri* showing fluted nature of warts. The larger of the suspensors is seen in the foreground.

Fig. 3. Part of a similar zygospore at higher magnification.

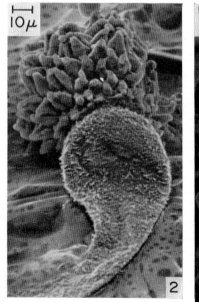
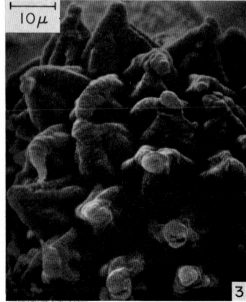

Plate II

Scanning electron micrographs. All scales equal 10μ.

Fig. 1. Conidia of a species of *Cunninghamella*, showing tapering spines with hemispherical bases.

Fig. 2. Part of the wall of the sporangium of *Mucor mucedo* showing spines of similar shape to those of the conidia of *Cunninghamella*.

Fig. 3. Group of ascospores of *Elaphomyces granulatus*.

Fig. 4. Ascospore of *E. granulatus* showing aggregation of rod-like thickenings to give wart-like sculpturing.

Fig. 5. Ascospore of same species in which constituent rods are more separated giving a more spiny appearance.

Fig. 6. *Hymenogaster hessei*. Young pyriform basidium at bottom of picture, showing two sterigmata, the one in the rear with developing spore. Mature detached spore at top of picture showing bottle-neck shaped base which fitted over sterigma of parent basidium.

Fig. 7. *H. hessei*. Basidiospore showing tubular basal part seen end-on.

12. SEM of Fungi and its Bearing on Classification

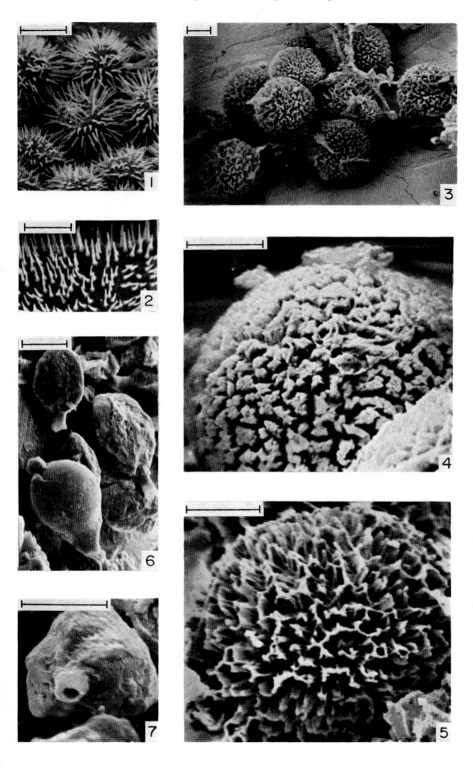

(e) Ascospores of *Tuber* spp. and some other members of the Tuberales and Pezizales. An examination of ascospores of several species of *Tuber* (the true truffle) with the SEM justifies the traditional inclusion of species having spiny ascospores and others with reticulately ornamented ones in the same genus (Fig. 2). It can be seen that the spines are connected at the base by ridges which

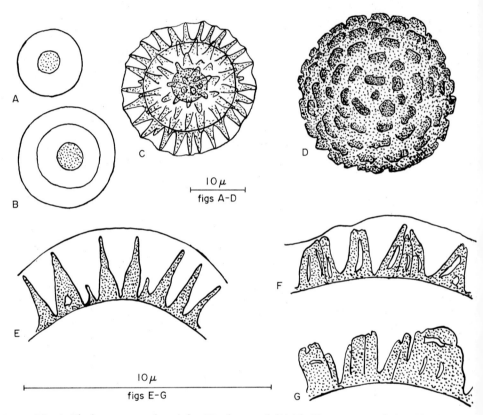

Fig. 1. *Elaphomyces granulatus* (after Hawker, et al. (1967). *Trans. Br. mycol. Soc.* **50**, 129–136). A. young hyaline, smooth-walled ascospore before discharge from ascus. The stippled area is the cell lumen. B. older spore after discharge from ascus, a peripheral layer or exospore is differentiated. C. older but still immature spore, pointed rods developed in exospore, pigmentation beginning. D. mature heavily pigmented spore with wart-like ornamentations. A–D as seen by light microscope. E. Section of epispore of spore at stage C immediately after collection. F. section of epispore of similar spore after drying, showing spines clumping together to form warts, and shrinkage of matrix. G. Section of epispore of mature spore, matrix dried down and no longer visible. E–G drawn from transmission electron micrographs.

show up best in partially collapsed spores. These ridges form a shallow reticulum and the spines are situated at the angles of this. In the more obviously reticulate spores (i.e. those described as such in literature dealing with the genus) the reticulations are deeper and the angles may or may not be extended to form a short spine.

Other genera of the Tuberales having spores usually described as "warty", such as *Genea* or *Hydnotrya*, are also worth studying with the SEM. Preliminary investigations suggest that details of spore ornamentation revealed by this instrument may assist in defining species. In such a difficult group as the Tuberales additional criteria of this kind are of great value in classification. Species of certain genera cannot be distinguished on one character alone but must be defined on a combination of characters (Hawker, 1954). Such a precise character as details of spore ornamentation is thus particularly useful.

A large convoluted fleshy fruit-body found growing just beneath the soil

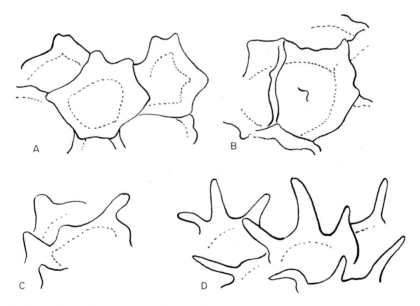

Fig. 2. Portions of ascospore surfaces of 3 species of *Tuber*. Sketched from scanning electron micrographs to approximately the same scale. A. *Tuber puberulum*. Reticulate thickening. Compartments with thick partitions, angles prolonged into blunt protuberances. B. *Tuber aestivum*. more delicate reticulations. C. *Tuber rufum*. Young spore showing developing spines. Shrinkage of spore during processing has left ridges between spines. D. *T. rufum* nearly mature slightly shrunken spore showing spines still connected by low ridges (which are not discernible by light microscopy).

surface in an Abingdon garden was thought by the collector to be a member of the Tuberales (such as *Geopora*, which it resembled in gross morphology) but details of the spore surface indicated that the specimen was an unusual form of a *Peziza* viz. *Peziza proteana* (Boud.) Seaver var. *sparassoides* (Boud.) Korf. Although the spore ornamentations were sufficiently obvious to be determined by light microscopy, the SEM showed them in greater detail (Hawker, 1968).

(f) Conidia of *Aspergillus* spp. Specific differences in ornamentation of conidia of several species of *Aspergillus* were demonstrated by Ellis and Heseltine (1969) by means of the SEM. Here is another example of a difficult genus where the instrument may be used to confirm or challenge existing classification and to provide further precise details as taxonomic criteria.

(g) "Smut" spores of the Ustilaginales. Banerjee *et al.* (1969) used both the scanning electron microscope and carbon replica techniques to study the smut spores (teleutospores or chlamydospores) of *Tilletia caries* (DC.) Tul. and concluded that the fine details of the exospore thus revealed offered a precise means of identifying the species of *Tilletia* and *Ustilago*. Attempts have previously been made to use the transmission electron microscope (Swinburne and Matthews, 1963) to study spore ornamentation in these economically important fungi, but the results did not resolve the problem of distinguishing species. The SEM is a much more suitable instrument for this purpose.

(h) *Hysterangium* spp. During a collecting excursion in the McCall region of Idaho, U.S.A., led by Dr. A. H. Smith of the University of Michigan, numerous collections of a species of *Hysterangium* were made by the writer and others. The basidiospores when seen under the light microscope showed faint markings but it was not until the SEM became available that these could be seen to be small hemispherical protuberances. A comparison of SEM pictures of basidiospores of several species of *Hysterangium* finally allowed the Idaho material to be assigned to *Hysterangium separabile* Zeller. (Hawker, 1969). In addition the SEM showed clearly the way in which the bottle-neck shaped basal part of the basidiospore fitted over the sterigma, a matter which is difficult to demonstrate by light microscopy. A similar arrangement occurs in other genera of hypogeous Gasteromycetes, such as *Hymenogaster* (Plate II, Figs 6 and 7).

2. *The Mode of Development of Fungus Spores and Its Bearing on Classification and Evolution.*

(a) Conidial scars. In recent attempts to devise a more natural classification of the Deuteromycetes considerable weight is given to the mode of origin of the conidia. The scar at the point where a conidium became detached from the

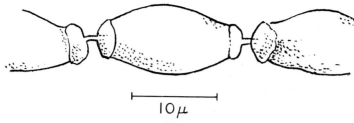

Fig. 3. *Sclerotinia fructigena*. Part of a chain of macroconidia drawn from a scanning electron micrograph (Willetts and Calonge (1969) *New. Phytol.* **68,** 123–131). Shrinkage during processing caused partial separation of spores, still attached by cytoplasmic thread.

parent structure can often provide useful evidence. Greenhalgh (1967) used the SEM in a study of the conidial scars of members of the Xylariaceae. C. V. Subramanian, while on a visit to my laboratory, used it to study the scars on conidia of species of *Drechslera* and obtained information of assistance to him in classifying these species (Subramanian, 1969). This is a field where the SEM could well be used to solve a number of difficult taxonomic problems.

Willetts and Calonge (1969) by observing slightly shrunken chains of conidia of *Sclerotina fructogena* Aderh & Ruhl, with the aid of the SEM demonstrated that the septal pores between the spores remained open until maturity and that cytoplasmic continuity was maintained (Fig. 3).

(b) Wall structure in spores impermeable to fixatives and embedding materials. In a study of the development of the zygospore of *Rhizopus sexualis* difficulty was experienced in fixing, embedding and sectioning the mature zygospores. The SEM had already been used in conjunction with the TEM to study the development of the warty mesospore and the disintegration of the membranous exospore (Hawker and Gooday, 1968). Fixation and embedding for observation with the TEM were, however, unsatisfactory with mature zygospores after the development of an impermeable endospore. Information relating to the form of this inner wall layer was obtained by fragmenting the spores by application of liquid nitrogen and examining the broken edges with the SEM. This was briefly reported, but further work is in progress and the method is being applied to the zygospores of a number of other species of the Mucorales. If the zygospores could be induced to germinate, the behaviour of the various wall layers could enable a comparison to be made with Hemiascomycetes and other fungi and perhaps throw some light on probable evolutionary pathways. Unfortunately germination of zygospores is not readily induced and so far we have been unsuccessful in finding a species in which germination occurs more readily.

CONCLUSIONS

The SEM is potentially of great value in the study of development of fungal structures and particularly of spores of various kinds. It has been used in a number of such studies but its use in other directions, such as the study of epiphytic or soil microorganisms *in situ*, while potentially valuable is still to be developed. While the instrument is invaluable for certain types of investigation it is best used in association with other methods. The results obtained may then be checked and compared with those obtained by light microscopy and by other techniques of electron microscopy (i.e. examination of ultrathin sections, carbon replicas and freeze etched preparations). Results with any one of these techniques are usually more easily interpreted in the light of results by other methods. It is also to be expected that the instrument itself will be modified to enlarge its scope and to increase definition and magnification and that the techniques of preparing and coating the specimens will also be improved. These developments will open up further possibilities and one can venture to prophesy that in the near future mycologists will find many uses for the SEM in addition to the ones I have indicated.

ACKNOWLEDGEMENTS

Thanks are due to Professor H. E. Hinton F.R.S. of the Department of Zoology, University of Bristol, for allowing me to use the Stereoscan microscope provided for him by the Science Research Council; to Professor C. V. Subramanian and Dr. H. J. Willetts for permission to refer to work done by them in the Department of Botany of Bristol University while they were visitors there, to my former research assistants Mrs. M. A. Gooday and Mrs. B. Thomas for their help in preparing material for viewing with the scanning microscope and to Miss M. Poole for preparing the photographs.

REFERENCES

BANERJEE, U. C., BANERJEE, S. and SAXENA, S. K. (1969). The transmission and scanning electron microscopy of some smut spores, with special reference to the chlamydospores of *Tilletia caries* (DC) Tul. XI *International Botanical Congress, Seattle: Abstracts* p. 8.

BERKELEY, M. J. and BROOME, C. E. (1837–1885). Notices of British Fungi. *Ann. Mag. nat. Hist.* Ser. 1, **6,** 430–39.

BERKELEY, M. J. and BROOME, C. E. (1846). Notices of British hypogeous fungi. *Ann. Mag. nat. Hist.* Ser. 1, **18,** 73–82.

ECKBLAD, F. E. (1962). Studies in the hypogaean fungi of Norway II. Revision of the genus *Elaphomyces*. *Nytt. Mag. Bot.* **9,** 199–209.

ELLIS, J. J. and HESELTINE, D. W. (1969). Surface configuration of spores in the *Rhizopus* and *Aspergillus flavus* groups. XI *International Botanical Congress, Seattle: Abstracts*, p. 54.

GRAY, T. R. G. (1967). Stereoscan electron microscopy of soil microorganisms. *Science N.Y.* **155,** 1668–70.

GREENHALGH, G. N. (1967). A note on the conidial scar in the Xylariaceae. *New Phytol.* **66**, 65–66.

HAWKER, L. E. (1954). British hypogeous fungi. *Phil. Trans. r. Soc. B.* **237**, 429–546.

HAWKER, L. E. (1968a). Wall ornamentation of ascospores of species of *Elaphomyces* as shown by the scanning electron microscope. *Trans. Br. mycol. Soc.* **51**, 493–498.

HAWKER, L. E. (1968b). *Peziza proteana* var. *sparassoides*. *Trans. Br. mycol. Soc.* **51**, 17–18.

HAWKER, L. E. (1969). A species of *Hysterangium* from Idaho attributed to *H. separabile*. *Mycologia* **61**, 115–119.

HAWKER, L. E. and GOODAY, M. A. (1968). Development of the zygospore wall in *Rhizopus sexualis* (Smith) Callen. *J. gen. Microbiol.* **54**, 13–20.

HAWKER, L. E., FRAYMOUTH, J. and DE LA TORRE, M. (1967). The identity of *Elaphomyces granulatus*. *Trans. Br. mycol. Soc.* **50**, 129–136.

HAWKER, L. E., THOMAS B. and BECKETT, A. (1970). An electron microscope study of structure and germination of conidia of *Cunninghamella elegans* Lendner. *J. gen. Microbiol* **60**, 181–189.

HOLLÓS, L. (1911). "Fungi Hypogei Hungariae", Budapest.

LANGE, M. (1956). Danish hypogeous macromycetes. *Dansk. bot. Ark.* **16**, 5–84.

SUBRAMANIAN, C. V. (1969). Patterns of conidial morphogenesis in *Drechslera* and their significance. XI. *International Botanical Congress, Seattle: Abstracts*, p. 212.

SWINBURNE, T. R. and MATTHEWS, H. I. (1963). The surface structure of chlamydospores of *Tilletia caries*. *Trans. Br. mycol. Soc.* **46**, 245–8.

VITTADINI, C. (1831). "Monographia Tuberacearum", Milan.

VITTADINI, C. (1842). "Monographia Lycoperdineum", Milan.

WILLETTS, H. J. (1968). Stromatal rind formation in the brown rot fungi. *J. gen. Microbiol.* **52**, 271–273.

WILLETTS, H. J. (1969). Structure of the outer surfaces of sclerotia of certain fungi. *Arch. Mikrobiol.* **69**, 48–53.

WILLETTS, H. J. and CALONGE, F. DE D. (1969). Spore development in the brown rot fungi (*Sclerotinia* spp,), *New Phytol.* **68**, 123–131.

YOUNG, T. W. K. (1968). Electron microscope study of the asexual structures of the Mucorales. *Proc. Linn. Soc. Lond.* **179**, 823–826.

13 | Étude Ornementale de Basidiospores au Microscope Électronique à Balayage

ROGER HEIM et JACQUELINE PERREAU

Laboratoire de Cryptogamie,
Muséum National d'Histoire Naturelle, Paris, France

Abstract: This paper shows how the Scanning Electron Microscope (SEM) has given increased information about the structure of spore ornamentation in Basidiomycetes.

The first section refers to the genus *Russula* and several species from tropical Africa have been studied, mostly from the section Pelliculariae. These species, which include some with a ring on the stipe, show considerable diversity of exosporial sculpturing when examined by the SEM (Cambridge "Stereoscan"), from the isolated warted type (verrucose and echinulate), through a mixed system of warts and ridges forming complete or incomplete reticulations, to a system almost entirely comprising a network of ridges. The ornamentation appears much more elaborate and complicated than can be made out by the ordinary light microscope; it is thus possible to define more accurately the real structure. The similarity with other Asterosporales, e.g. *Lactarius* (epigeous) and the Asterogastrae (hypogeous), is increased.

Because of the great importance of sporal characters in discussions of the systematics of the Gasteromycetes, spores of the following groups have also been examined by the SEM: Tulostomataceae, Phellorinaceae and Lycoperdaceae. The second section of the paper presents results and interpretation of the electron micrographs obtained and comparison is made with illustrated data from light microscopy. About forty species belonging to twelve genera have been examined. The spores prove to be quite or nearly globose and mostly of small size. Details of the ornamentations are described: pillars in *Geastrum* and *Trichaster*, spines in *Queletia*, warts and ridges in *Tulostoma*, a catenulate incomplete network in *Lycoperdon echinatum* and a spinulose one in *Calvatia lilacina*. It is established that the spores of *Phellorina* and *Dictyocephalos* (*Whetstonia*) do not show a reticulum but only crowded tuberculose warts, whereas those of *Mycenastrum* are not just verrucose but are covered with a complete network which is often thick giving the spores a foveolate (or dimpled) appearance. The disposition of ornamentation especially around the hilar appendix, where it often shows a stellate arrangement, is also indicated. The relationships of the various kinds of ornamentation are again pointed out. Among the Gasteromycetes, the SEM reveals, in the spores, a scarcity of isolated, simple ornaments and a great prevalence of spores with incomplete networks.

I. Sur Quelques Russules Tropicales Africaines

INTRODUCTION

L'existence sur les spores de plusieurs groupes de Champignons mais particulièrement chez les Astérosporales—aussi bien gastéroïdes et hypogées qu'agaricées—d'une ornementation périsporique et amyloïde inégalement sus-jacente au relief exosporique, a suscité de nombreux travaux (G. Malençon, R. Heim, M. Josserand, M. Locquin, J. Perreau). Ils se sont appliqués à la description de ces ornements, soit d'un point de vue systématique en recherchant la valeur spécifique de leur agencement décoratif, soit en tentant d'expliquer la répartition de la périspore par l'analyse de la structure tégumentaire et de sa différenciation au cours de la croissance sporale, soit enfin en s'efforçant de dégager la signification phylétique qui peut correspondre au sens évolutif des motifs ornementaux précisés. En fait, ces soucis se rejoignent dans un but unique, celui de la connaissance sans cesse plus approfondie de l'ornementation. La microscopie à balayage est naturellement venue apporter son appui à ces investigations, mais alors la distinction entre exospore et périspore n'apparaît plus ainsi que le montrent des documents récemment présentés pour des Lactaires et des Russules (Perreau et Heim, 1969). Ce sont les divers aspects de la décoration sporale de quelques Russules Nigricantinae, Heterophyllae et surtout Pelliculariae (Discopodinae et Radicantes), leur comparaison avec les résultats obtenus en microscopie photonique et le parti qu'en peut tirer la systématique, qui seront envisagés ci-après.

De cette étude descriptive sur la décoration sporale résulte un classement propre aux motifs ornementaux correspondants. Il ne saurait revêtir, bien entendu, qu'une signification partielle, appliquée à un ensemble systématique limité et relativement homogène.

ORNEMENTATION VERRUCO-RÉTICULÉE

1. Ornementation Verruco-réticulée et Ailée, Parfois à Réseau d'Alvéoles Entièrement Fermés

Russula carbonaria Heim et Gilles, ad. int. (no. Ag.G.44)* (Fig. 1a et Planche I, Figs 1 et 2).

Le *Russula carbonaria*, miniature de Nigricantinae, à chair entièrement noircissante, possède une spore à ornementation à la fois tuberculeuse et réticulée du type alvéolé-ailé, mais les alvéoles, polygonaux ou grossièrement qua-

---

* Abréviations: G = leg. G. Gilles; H = R. Heim
    PHM = microscope photonique ou optique
    SEM = microscope à balayage

drangulaires, sont parfois entiers (Planche I, Fig. 2); quelques "bracelets"* de jonction, annulaires ou carrés, s'y montrent çà et là de même que des "grilles" et "treillis" incomplets; autrement dit, l'ornementation rappelle de près celle de certains Lactaires. Les photographies au PHM et les dessins correspondant à cet examen traduisent également la diversité, l'hétérogénéité, la variabilité de la décoration, ses mailles et crêtes souvent appendiculées ou interrompues-lovées; quelques bracelets se localisent sur celles-ci. Le SEM révèle des dimensions sporales légèrement mais nettement inférieures à celles vues au PHM.

Dans l'ensemble, l'ornementation apparaît enchevêtrée et quelque peu confuse, mais le SEM révèle parfaitement le détail, notamment des crêtes sinuées, peu festonnées, et des courts aiguillons isolés.

2. Ornementation Verruco-réticulée et Ailée aux Alvéoles Parfois Entièrement Fermés, Mais aux Verrues Prédominantes

Russula annulata Heim (type = f. *violacea* var. *molochina* Heim, ad. int. (Fig. 1b; Planche I, Fig. 3), *annulata* s. sp. *parasitica* Heim (H. G. no. 5001) (Fig. 1c; Planche I, Fig. 4a, b), *cyanea* Heim (Fig. 1d; no. Ag. G. 47) (Planche I, Figs 5, 6a, b), *carmesina* Heim (H, LM 3038) (Fig. 2b; Planche II, Fig. 4).

Nous avons examiné au PHM et au SEM ces quatre Russules Pelliculariae du groupe des Discopodinae (parmi les Heliochromae) au stipe adhérant au support par un disque—les 3 dernières, mineures—se différenciant par le pigment (respectivement lilas, rouge, bleu, carmin). Elles offrent une spore sphérique dont l'ornementation est en partie réticulée, accompagnée de tubercules cylindracés ou coniques, moins prédominants que les épines digitées dans le type *annulata* (Perreau et Heim, 1969, Planche IV, Fig. 1).

Les différences dans les spores de ces espèces affines s'appliquent tout d'abord aux dimensions précises de ces organes, un peu plus grandes dans *cyanea*, aux alvéoles profonds, que dans *parasitica* où le réseau est très marqué et les verrues cylindriques, aux fort aiguillons plus prépondérants dans *molochina*, au réticulum moins complet dans *carmesina*.

Sur ces Russules, le PHM traduit avec plus de netteté l'importance du réseau que le SEM qui met en évidence surtout le détail des alvéoles et leur profondeur, mais les éléments ornementaux secondaires y sont bien mieux précisés. Ainsi, la quasi totalité des dessins sporaux rapproche intimement les 4 espèces, moins

---

* Nous avons appelé "bracelets" les petites figures, assez fréquentes sur les spores de certains Lactario-Russulés, qui constituent le minuscule périmètre de jonction de diverses lignes amyloïdes autour du noeud théorique vers lequel convergent ces connectifs (Heim, 1955). Pour les grilles et treillis, même référence.

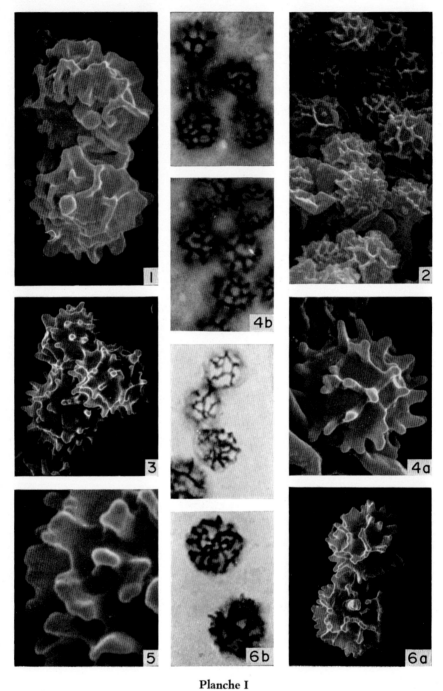

Planche I

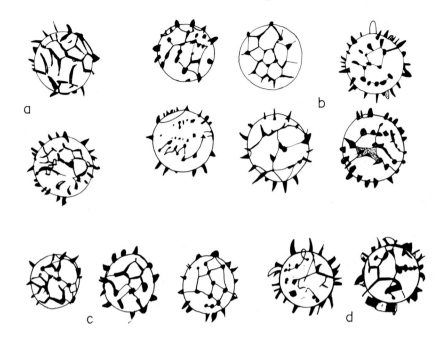

Fig. 1. Spores observées dans le réactif de "Melzer" de: a. *Russula carbonaria*; b. *R. annulata* var. *molochina*; c. *R. annulata* s. sp. *parasitica*; d. *R. cyanea*.
×2000.

Planche I

Fig. 1. *Russula carbonaria*, spores réticulées-tuberculeuses du type alvéolé-ailé.
SEM, × *ca* 5300.
Fig. 2. *R. carbonaria*, groupe de spores offrant différents aspects de l'ornementation.
SEM, × *ca* 2100.
Fig. 3. *R. annulata* var. *molochina*, spores réticulées à forts aiguillons.
SEM, × *ca* 2700.
Fig. 4a. *R. annulata* s. sp. *parasitica*, spore réticulée à tubercules cylindriques.
SEM, × *ca* 6600.
Fig. 4b. *R. annulata* s. sp. *parasitica*.
PHM, × *ca* 1500.
Fig. 5. *R. cyanea*, détail d'une spore.
SEM, × *ca* 10,000.
Fig. 6a. *R. cyanea*, spores réticulées, à profonds alvéoles.
SEM, × *ca* 2700.
Fig. 6b. *R. cyanea*, spores jeunes ou parvenues à maturité.
PHM, × *ca* 1500.

Clichés Laboratoire de Géologie du Muséum, Paris et R. Heim (Fig. 4b et 6b).

bien la dernière, en appuyant une parenté que pourrait rejeter a priori la diversité des caractères morphologiques et surtout pigmentaires, bien différents. Autrement dit, le rapprochement taxinomique que l'un de nous a précédemment proposé trouve ici, dans le détail semblable de l'architecture tégumentaire de la spore, un appui décisif. Taille des carpophores, habitat et adaptation biotique, couleurs, appartiennent à des variations secondaires. Les indices sporaux s'ajoutent au contraire à la minceur du pileus, à la coloration des spores, à la saveur douce de la chair et à d'autres critères anatomiques. Le *R. carmesina* cependant s'éloigne de ces espèces par la présence des bracelets carrés dans l'ornementation sporale et par le revêtement piléique.

Nous nous étendrons plus spécialement sur l'analyse d'une telle configuration dans l'une de ces espèces, *Russula cyanea*.

L'étude sporale de cette Russule, minuscule Pellicullariae, livre une architecture particulièrement intéressante: elle montre un système complexe et très apparent, très profond, dans des spores sphériques qui, jeunes, révèlent un dessin subtil mais précis, superposable à celui des spores mûres. Le réseau est à la fois bien formé et à mailles inégales, marqué aux noeuds de tubercules aigus proéminents. Le PHM en donne une image précise, le SEM en creuse le détail, livrant les précisions propres à l'insertion des tubercules sur les crêtes du réseau. Le profil de ces dernières complète le connaissance de l'ensemble ornemental;

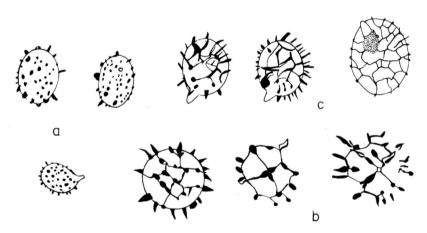

Fig. 2. Spores observées dans le réactif de "Melzer" de: a. *Russula acutispora*; b. *R. carmesina*; c. *R. discopus*.
   ×2000 sauf c, à droite, ×2600.

l'homogénéité de celui-ci rend la vision claire et rigoureuse tout en approfondissant les accidents du réticulum sur les crêtes: pustules, nodules ou bosses, échinulations cylindriques aux extrémités arrondies et non renflées, flancs des alvéoles. Ici la combinaison des deux modes d'examen conduit à un enregistrement complet de la décoration.

On serait certes tenté de rapprocher également ces espèces du *Russula carbonaria*, signalé ci-dessus (Planche I, Figs 1, 2), dont le type d'ornementation sporale présente des analogies apparemment étroites avec le précédent: forme et coexistence d'un réseau partiel et d'échinulations ou de digitations aplaties, plus épaisses, mais l'examen approfondi des ornements du *carbonaria*, aux caractères macroscopiques et anatomiques d'ailleurs très différents, révèle la présence de bracelets, indice qui n'existe pas ailleurs sauf dans *carmesina*, et surtout d'une haute variabilité selon les spores.

3. Ornementation Verruco-réticulée et Ailée Hétérogène

Russula discopus Heim (no. H, A.O.F., D 38) (Fig. 2c; Planche II, Fig. 5).

Le cas du *R. discopus* justifie une place privilégiée. Mineure, lignicole, discopode, au pigment jaune citron, à lames distantes et très fragiles, exannulé mais indiscutablement solidaire des Annulatae, ce champignon apporte cependant l'exemple d'une ornementation sporale en grande partie fortement échinulée, aux aiguillons longs et pointus vus au PHM, s'ajoutant à des connectifs très variables, parfois délicats, linéaires et parallèles (treillis), parfois simples, joignant deux tubercules, plus rarement constituant un réseau presque complet. Une tache hilaire amyloïde très visible encapuchonne parfois la partie inférieure et moyenne de l'appendice hilaire. De cette diversité le SEM donne l'image d'une architecture mixte, à la fois confuse et très émergente, où les éléments réticulés disparaissent en grande partie parmi les ornements verruqueux érigés: cylindracés, étirés, noduleux, ou difformes, soudés, assemblés. Mais la variabilité et l'hétérogénéité de l'ornementation vue à la fois à l'un et à l'autre microscopes si elles en donnent une image très compliquée et notablement éloignée de celle des autres Discopodae ne peuvent suffire à négliger les particularités physionomiques du groupe, notamment le disque basal du pied, l'extrême fragilité et la minceur de la chair, les sillons piléiques profonds, et, quant aux spores, l'aptitude à livrer l'une ou l'autre des deux constructions réticulée—en général partiellement—ou échinulée sur le même élément ou à la fois les deux modes—soit le type verruco-réticulé. Ajoutons que la différenciation des ornements dans cette espèce se fait par la voie destructive, contrairement au cas de *R. cyanea*, à partir de nodules isolés et d'une pellicule générale amyloïde aussi bien que par la voie constructive par formation précoce et autoprogression du réseau amyloïde.

C'est ainsi que cet exemple met en évidence la valeur d'une confrontation entre les caractères sporaux et physionomiques devant les tentations auxquelles peut succomber le mycologue à la recherche des affinités naturelles.

Pour conclure, l'examen comparatif des ornementations sporales de Russules Pelliculariae appartenant à la stirpe *annulata- parasitica, cyanea, typica-molochina*— à réseau amyloïde, spore sphérique et disque pédiculaire basal, et du *R. discopus* à ornementation mixte trouve des arguments confirmatoires par l'examen au SEM. Ces trois premières espèces nous ont offert des spores dont l'architecture est très voisine, pratiquement superposable, la présence des aiguillons ne s'y révélant qu'additive, tandis que la 4ème espèce transmet une complexité décorative accusée, désordonnée, du fait de la nature hétérogène des ornements et de la prédominance fréquente des épines. Rappelons que dans le type *annulata*, ces dernières, digitées, sont plutôt dominantes, trait d'union qui conduit au rapprochement des quatre espèces composant ce groupe.

ORNEMENTATION ÉCHINULÉE

1. *Ornementation à Échinulations Cylindro¹des, Serrées*

Russula acutispora Heim, ad. int. (no. Ag. G. 41) (Fig. 2a; Planche II, Fig. 3), *R. alveolata* Heim (no. H., LM. 3060) (Planche III, Fig. 6), *R. africana* Heim, ad. int. (no. H., LM. 1334) (Planche II, Figs 1, 2).

Dans le *Russula acutispora*, espèce pelliculaire, l'appendice hilaire, long, grêle, aigu, de 1·8 à 2·3 μ, et les échinulations isolées et inégalement serrées, pontiformes, coniques, nodulaires, ou étroitement cylindroïdes, de 0·6 à 0·7 μ, voire 1·1 μ de haut, traduisent l'originalité de l'ornementation: la variabilité et la diversité de ces appendices, leur faible coloration amyloïde, la très visible autoformation par prolifération constructive à partir de la pellicule périsporique iodophile, s'ajoutent à l'absence totale de connectifs et à la révélation des petits ornements sous-jacents. Tous ces caractères sont parfaitement mis en évidence par le SEM qui accuse ces détails, notamment ce dernier. Comme il convient généralement, les spores vues par cet examen sont un peu plus petites que sous le PHM.

Le *Russula alveolata* possède une ornementation sporale très proche du précédent. Espèce de taille moyenne du groupe des Pelliculariae, il présente un hyménium alvéolé très prononcé et de grandes spores de l'ordre de 10 × 8 μ (o.i.) vues au PHM, un peu plus petites au SEM. Elles sont couvertes de longs aiguillons cylindracés un peu coniques, de 0·9–1·2 μ de haut, à assise dilatée, qui atteignent 1·6–1·9 μ de haut au SEM. Indépendants, ces piliers cylindroïdes s'élargissent en arcs-boutants à leur base, en continuité avec l'exospore sousjacente dont l'assise est pratiquement lisse, marquée seulement de rares et petites nodosités. Ainsi, pour cette espèce, la seule distinction dans les deux examens microscopiques respectifs concerne la forme entièrement cylindrique des épines digitées, arrondies au sommet vues au SEM et non pas tronconiques, planes et coupantes à l'extrémité, observées au PHM.

Dans le *R. africana*, champignon centrafricain, mais probablement très répandu en Afrique tropicale, proche de l'*alveolata*, mais à lames crème, décurrentes et non alvéolées, et à spores plus petites, nous retrouvons le type d'ornementation sporale à échinulations cylindroïdes identiques à celles de la précédente espèce, rarement aplaties mais un peu plus hautes, groupées en formation doublement digitée.

2. Ornementation à Échinulations Cylindroïdes ou Tronconiques Distantes

Russula mimetica Heim (no. H.G. 5001 bis) (Fig. 3c; Planche III, Figs 5a, b).

Dans cette espèce gabonaise, minuscule, sosie de la s.-sp. *parasitica*, comme celle-ci munie d'un anneau, simple ou double, et à pigment rouge, on est en présence d'un représentant pelliculaire de la section des Radicantes, à spore obovoïde ponctuée de courts aiguillons tronconiques indépendants, de moins de 1 μ de haut, souvent alignés. L'examen au SEM s'accorde bien avec la vision au PHM, mettant parfaitement en exergue la fréquente ordonnance en lignes de ces verrues, une relative régularité de leur diamètre, une taille comparable à celle relevée au PHM. Cependant la base de ces ornements (dont le diamètre du col atteint 0·4 μ) vus au SEM s'élargit généralement en forte assise pouvant dépasser 1·5 μ de diamètre, attachée au tégument sporique et en continuité avec lui, tandis que le sommet se dilate parfois selon une coiffe subsphérique de 0·50–0·55 μ, l'ensemble de l'ornement mimant l'aspect d'une quille.

Ainsi se confirme l'existence de deux groupes au moins de Russules annelées bien différentes, séparées par l'habitat, l'adaptation biotique et l'ornementation sporale, dont l'un de nous a fait respectivement les Discopodinae, parmi les Heliochromae, et les Radicantes parmi les Aureotactae (Heim, 1938).

Russula echinosperma Heim et Gilles, ad. int. (no. Ag. G. 46) (Planche III, Fig. 3a, b).

Dans le *Russula echinosperma*, les spores sont caractérisées par une asymétrie accusée, la forme amygdaloïde, un peu fusiforme en profil frontal, exagérée par l'étirement du système hilaire et l'existence d'une plage supra-apiculaire correspondante très développée, enfin d'échinulations isolées, distantes, apparaissant au PHM comme des aiguillons fins et dressés de 0·8–1 μ de haut environ. L'observation au SEM rend très apparents ces divers dispositifs sur des spores

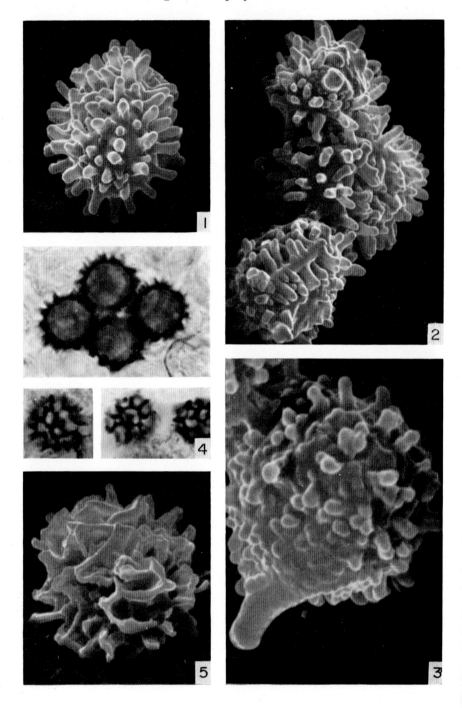

Planche II

nettement réduites de 1/9 environ. Les ornements, érigés, droits, parallèles, privés d'anastomoses, atteignent 1·6 μ de haut, s'amincissant insensiblement vers le sommet. Sur le soubassement exosporique se manifestent quelques pustules de petite taille et quelques rides indécelables au PHM, surtout abondantes au voisinage de l'insertion hilaire, à la périphérie de la plage. Aucun indice d'anastomose ne se révèle. Cet aspect extrême correspond au type sporal à ornementation strictement aiguillonnée—subtronconique, l'indépendance individuelle des épines, assez distantes, restant entière.

3. *Ornementation constituée de Verrues Basses, de Bosses, Généralement Simples et Distantes, et de Rares Connectifs*

Russula papillata Heim et Gilles, ad. int. (no. Ag. G. 34) (Fig. 3b; Planche III, Fig. 2).

Le *Russula papillata*, à spores brièvement ovoïdes, offre au PHM des verrues bien distantes, donc relativement peu nombreuses, arrondies et basses, ou rarement tronconiques, réunies parfois par de brefs connectifs. Au SEM, les spores apparaissent plus petites (6·1 contre 6·6 μ p. ex., ornements inclus) et comportent des verrues nettement plus grosses (0·7 à 1 μ en général) auxquelles s'ajoutent çà et là en arrière-plan des bosses confluentes formant un soubassement au pavage presque continu. L'appendice hilaire, conique, est très développé et la plage suprahilaire lisse. Ainsi, les spores vues au SEM mettent bien en relief les verrues dominantes et les deux niveaux d'ornementation, l'un supérieur, disjoint, aux tubercules cylindroïdes arrondis mais non élargis au sommet, l'autre inférieur, aux bourrelets peu émergents et tangents et aux très brefs et délicats ponts de connexion. Il n'y a pas contradiction avec la composition vue au PHM, mais confirmation pour les ornements, non déformés, et les très rares connectifs, et

---

Planche II

Fig. 1. *Russula africana*, spore à échinulations cylindroïdes.
SEM, × ca 7000.
Fig. 2. *R. africana*, différents aspects de l'ornementation.
SEM, × ca 5500.
Fig. 3. *R. acutispora*, face subdorsale d'une spore montrant l'appendice hilaire prédominant.
SEM, × ca 10,000.
Fig. 4. *R. carmesina*.
PHM × ca 1500.
Fig. 5. *R. discopus*, spore offrant une ornementation hétérogène.
SEM, × ca 6000.
Clichés Laboratoire de Géologie du Muséum, Paris et R. Heim (Fig. 4)

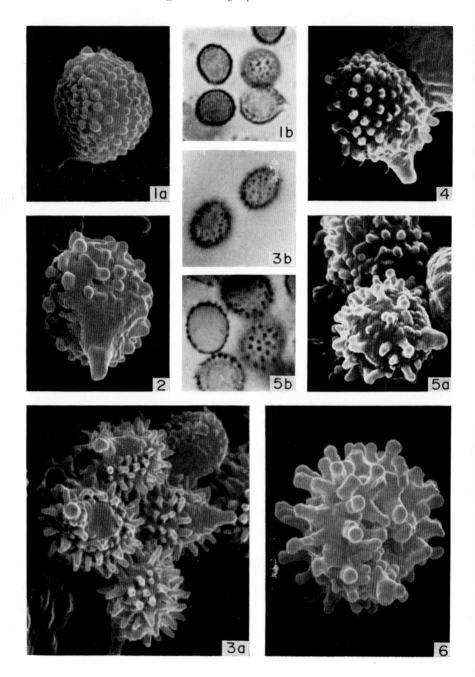

Planche III

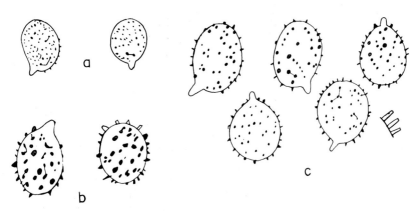

Fig. 3. Spores observées dans le réactif de "Melzer" de : a. *Russula viridicens*; b. *R. papillata*; c. *R. mimetica*.
× 2000 sauf c, à droite, × 2600.

Planche III

Fig. 1a. *Russula viridicens*, spore ornée de petites verrues arrondies.
    SEM, × *ca* 5000.
Fig. 1b. *R. viridicens*.
    PHM, ×ca 1500.
Fig. 2. *R. papillata*, spore à verrues et bosses inégalement distantes.
    SEM, × *ca* 5300.
Fig. 3a. *R. echinosperma*, spores à ornementation aiguillonnée.
    SEM, × *ca* 4500.
Fig. 3b. *R. echinosperma*, spores ornées d'aiguillons.
    PHM, ×ca 1500.
Fig. 4. *R. mimetica*, spore montrant de courts aiguillons tronconiques.
    SEM, × *ca* 5200.
Fig. 5a. *R. mimetica*, spores montrant des ornements à sommet dilaté.
    SEM, × *ca* 5200.
Fig. 5b. *R. mimetica*.
    PHM, ×*ca* 1500.
Fig. 6. *R. alveolata*, spore couverte d'épines digitées.
    SEM, × *ca* 5700.

    Clichés Laboratoire de Géologie du Muséum, Paris et R. Heim (Fig. 1b, 3b, 5b)

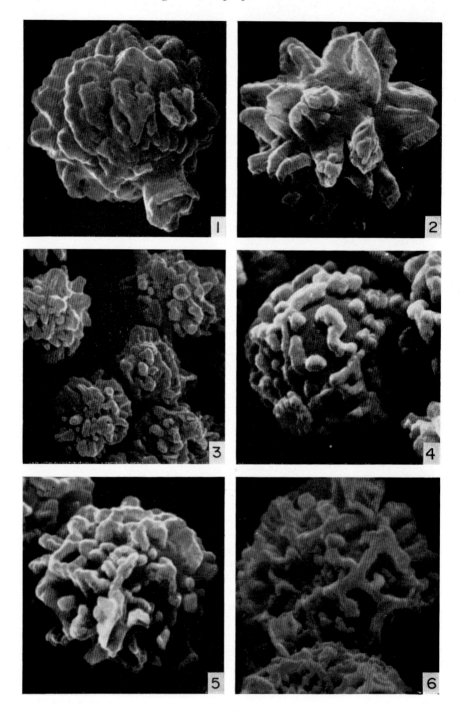

Planche IV

enfin révélation pour la nappe gondolée de base. Contrairement au *R. viridicens*, les premiers se montrent encore distants mais inégalement, le phénomène de coalescence n'apparaissant pas malgré l'action conjointe de la réduction du volume sporal et du grossissement des verrues.

Russula viridicens Heim et Gilles, *ad. int.* (no. Ag. G. 33) (Fig. 3a; Planche III, Fig. 1a, b).

Le *Russula viridicens*, entité tropicale parmi les Heterophyllae, proche du *cyanoxantha* et surtout de sa variété *variata* Sing., caractérise un type sporal brièvement ovoïde, mais nettement asymétrique, à très petites verrues qui, au PHM, se montrent aiguës, indépendantes et peu serrées, subéquidistantes mais de dimensions, surtout de largeur, variables. Le dessinateur écrirait : à verrues fines, nombreuses, isolées et basses; mais l'examen plus attentif révèlera la variabilité de taille de celles-ci et l'existence rare mais constante sur chaque spore de grêles connectifs rectilignes réunissant 2 ou 3 d'entre elles. Vue au SEM, la spore apparaît plus petite, réduite de 1/12 environ (de 6 à 5·5 μ p. ex.) alors que les ornements sont au maximum 2·2 fois plus gros qu'au PHM : ils sont devenus noduleux, serrés, également répartis en raison de la réduction des espaces, de volume non seulement inégal mais plus ou moins irrégulier, rarement parfaitement globuleux, parfois allongé ou allantoïde, voire se prolongeant en arc-boutant; parfois il y a jumelage de 2 tubercules au lieu d'anastomose et les connectifs apparaissent dans la plage supra-hilaire. Ainsi, la figuration au SEM traduit l'exagération dans l'importance des ornements et de leur variabilité morphologique; elle montre l'appartenance des brefs connectifs au système

---

Planche IV

Fig. 1. *Queletia mirabilis*, spore ornée de crêtes tuberculeuses-verruqueuses.
SEM, × *ca* 10,000.

Fig. 2. *Q. mirabilis*, ornementation formée de hautes épines, parfois coalescentes.
SEM, × *ca* 9000.

Fig. 3. *Tulostoma floridanum*, spores couvertes d'épines.
SEM, × *ca* 5000.

Fig. 4. *T. chevalieri*, spore montrant une ornementation composée de verrues et de crêtes arrondies et sinueuses ainsi que d'un anneau périhilaire.
SEM, × *ca* 10,500.

Fig. 5. *T. barlae*, tubercules et crêtes anastomosées sur une spore en profil dorsi-ventral.
SEM, × *ca* 10,000.

Fig. 6. *T. berkeleyii*, ornementation sporale interrupto-réticulée.
SEM, × *ca* 10,000.

Clichés Laboratoire de Géologie du Muséum, Paris

tuberculaire, mais elle ne modifie pas les qualités essentielles de la composition même des ornements : elle l'éclaire malgré l'effet combiné de la réduction du volume sporal et du grossissement des verrues prédominantes, noduleuses pour la plupart ou exceptionnellement étirées.

CONCLUSION

Cette étude d'un ensemble de Russules, toutes recueillies dans la forêt hygrophile d'Afrique intertropicale—Côte d'Ivoire, Oubangui, Gabon, met en évidence tout d'abord l'abondance relative des espèces de la section Pelliculariae et l'existence de formes annelées se rattachant à ces dernières. Cette investigation vient donc appuyer et développer les travaux d'ordre systématique, déjà publiés ou actuellement sous presse, se rapportant aux mêmes lieux et pareilles entités. Elle confirme et étend d'autre part de précédents examens entrepris au SEM sur l'ornementation sporale, appliqués entre autres aux Lactario-Russulés. Elle montre que celle-ci, examinée à ce microscope, comprend comme pour les Russules des zones tempérées les types réticulés (à alvéoles complets ou partiels, ou mixtes, c'est-à-dire verruco-réticulés) et échinulés avec les combinaisons variées des deux dispositifs extrêmes.

Au SEM, les dimensions des spores sont affectées en général d'une réduction, variable selon les espèces, et qui peut atteindre plus de 1/10, tandis que des détails se révèlent ou s'y montrent plus complets : ils concernent certaines unités ornementales, notamment celles du revêtement exosporique du corps sporal (*R. acutispora*) et la composition des crêtes (*R. molochina*). Les papilles ou les nodosités que recèle la surface basale du tégument, parfois minuscules, apparaissent notamment sur la plage hilaire (*R. echinosperma*), alors que ces protubérances de faible épaisseur échappent à l'observation au PHM comme dans *R. papillata* où ces ornements deviennent notablement grossis (ce qui se produit pour bien d'autres champignons dont la spore semble au PHM faussement lisse, comme dans *Pseudogomphus* Heim). D'une façon générale au SEM, les ornements s'allongent, s'épaississent, ou se dilatent dans leur partie terminale par rapport au PHM ; il se produit un émoussement des crêtes et des pointes, celles-ci devenant obtuses ; mais les connectifs les plus délicats subsistent souvent (*parasitica, viridicens*).

Ainsi l'architecture extérieure est légèrement exagérée ou modifiée et la révélation de la vérité s'affirme par la comparaison et l'addition des résultats obtenus respectivement aux deux microscopes. Cependant, en définitive, sans émettre une règle uniformément applicable à l'ensemble systématique en cause, on peut considérer que le SEM conduit à des figurations plus complètes et par conséquent plus proches de la vérité que le PHM.

Ajoutons, quoiqu'ils soient quelque peu en dehors de l'objet de cette Note, que les rapports entre la décoration sporale et la classification générale des Russules peuvent apparaître étroits dans les limites de la stirpe ou du groupe sans qu'aucun impératif d'affinité cependant puisse autoriser à en franchir l'affirmation. L'argument, positif ou négatif, tiré de la similitude ou de la distinction entre deux types fondamentaux aura ou non une valeur selon l'exemple dont il s'agit. Autrement dit, nous retrouvons, notablement grandi par la vision au SEM, l'intérêt du détail descriptif sans être certain *a priori* qu'on puisse l'étendre à des considérations de parenté.

Cependant, une conclusion est possible : l'exemple des deux Russules pelliculaires annelées, appartenant à deux stirpes physionomiquement proches, mais anatomiquement distinctes—*parasitica* (*annulata*) et *mimetica* (*radicans*), toutes deux pseudoangiocarpes, également supposées surévoluées ou dégradées, mais à décoration sporale toute différente, le prouve. Il appuie à l'égard de ces deux groupes la notion de convergence.

Ce précédent cas montre d'autre part qu'il n'est pas possible d'affirmer que le dispositif réticulé sporal corresponde à un état d'évolution progressive supérieur à celui qui ne comporte que des verrues isolées.

Ainsi, l'examen approfondi de l'ornementation basidiosporale des Russules en général n'ajoute rien de véritablement inédit à l'inventaire déjà dressé des modes de décorations telles que les auteurs les ont décrites, telles que nous les avons précisées et figurées au SEM pour divers genres d'Hyménomycètes, résultats que le présent travail s'est efforcé de développer. Mais elle conduit à donner des motifs architecturaux de la spore leur rigoureux agencement et les vraies limites de leurs variations.

II. A Propos De l'Ornementation Sporale des Gastéromycètes

Dans l'étude des Gastéromycètes et en particulier des Gastéromycètes "vrais", non agaricoïdes, auxquels nous nous limiterons ici, les caractères sporaux : forme, dimensions, ornementation surtout, jouent un rôle marquant ; sans être à eux seuls décisifs, ils apportent cependant des indications de grande valeur, en spécification notamment, pour des exemplaires à maturité où peu de critères demeurent utilisables. Or, les spores de ces Champignons sont bien souvent de petite taille et leurs ornements non définissables avec exactitude. Afin d'en mieux connaître les détails, nous avons examiné, au PHM, puis au SEM, les spores d'une quarantaine d'espèces réparties en douze genres, appartenant, selon la nomenclature de Malençon (1955) adoptée par l'un de nous (Heim, 1969), aux familles des Tulostomataceae, des Phellorinaceae et des Lycoperdaceae.

La Fig. 4 (a–h) offre quelques exemples du relief sporal tel qu'il peut être déchiffré habituellement au PHM ; les volumes et leurs contours se distinguent dans leur signification géométrique générale, mais ne livrent aucune précision ; ils apparaissent arrondis ou polygonaux, plus ou moins élancés tandis que leur enchevêtrement reste difficilement interprétable. Evidemment, lors des observations au SEM des mêmes spores, nous constatons une fois de plus que les ornements y sont plus nombreux puisque des plages semblant unies sont découvertes comme ornées. Non seulement la densité des aspérités se révèle supérieure à celle estimée au PHM, mais également leur accentuation par rapport à l'élément sporal tout entier. Ce phénomène correspond à une rétraction de ce dernier provoquée par l'intense dessiccation qu'il subit ; il en résulte un approfondissement des sillons séparant les ornements qui paraissent alors plus élevés. Au contraire, le montage des spores dans un liquide a pour effet de les dilater et par là-même d'en atténuer les inégalités superficielles. Pour les spores d'une espèce, il existe donc une différence appréciable, assez importante quoique peu aisément calculable, entre les mesures effectuées avec l'une ou l'autre méthode ainsi qu'un resserrement ou un espacement de la décoration.

Nous voyons, d'après les microphotographies au SEM des Planches IV à IX, que les ornements sporaux des Gastéromycètes envisagés, bien que très variables selon les espèces et les familles, se regroupent en trois types architecturaux, d'ailleurs liés entre eux, ainsi que nous l'avons déjà démontré pour des Agaricales et des Boletales (Perreau, 1967b ; Perreau et Heim, 1969). Les protubérances de plus faible amplitude, $0\cdot2$–$0\cdot3$ μ environ, ont été trouvées à la surface de la spore ovoïde du *Bovista plumbea* Pers. (M.N.H.N.P. no. 308 – Upsal. no. 1608) (Fig. 4g et Planche IX, Fig. 3). Ce sont de fines pustules et verrucosités qui semblent se propager en bosselures sur le long pédicelle ; elles apparaissent plus éparses et uniformes dans leur taille chez le *Lycoperdon pyriforme* Pers. (Exp. Myc., Lab. Crypt., 1969) dont Gregory et Nixon avaient soupçonné l'ornementation en 1950. Des verrues ou des tubercules à base circulaire et sommet hémisphérique, légèrement cylindriques toutefois, ornent des spores telles que celles du *Lycoperdon umbrinum* Pers. (M.N.H.N.P., Herb. Lloyd no. 5569) (Planche IX, Fig. 4), de Tulostomes comme *Tulostoma chevalieri* Har. & Pat. (M.N.H.N.P., no. 332 ; Dahomey, 1902) (Planche IV, Fig. 4) ou *T. campestre* Morgan (M.N.H.N.P., no. 332), de Geasters : *G. mirabile* (M.N.H.N.P., Herb. Lloyd), de l'espèce ou forme voisine, *G. subiculosum* Cooke & Massee (R.H., L.M. 3086b) (Planche VI, Fig. 1). Le sommet de ces verrues est parfois aplati ; un exemple en est donné par les spores du *Lycoperdon echinatum* Pers. (Perreau, 1967) (Planche IX, Figs 5 and 6) pour lequel cet aspect se devinait déjà au PHM (Fig. 4h). Pourtant le volume des protubérances est rarement régulier et l'on observe plus fréquemment des tubercules triangulaires ou polygonaux, chez les

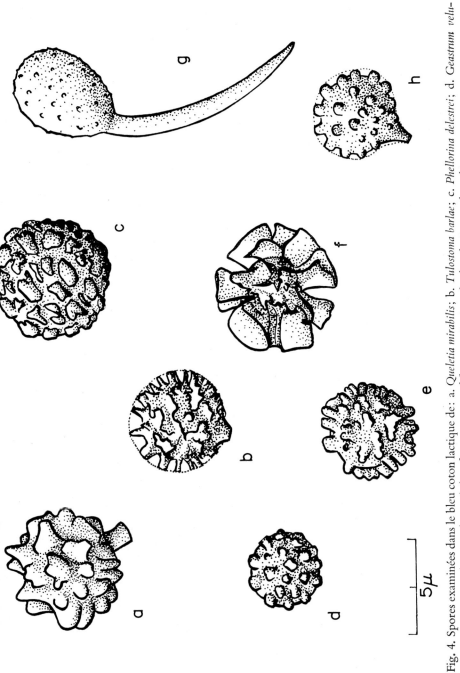

Fig. 4. Spores examinées dans le bleu coton lactique de : a. *Queletia mirabilis*; b. *Tulostoma barlae*; c. *Phellorina delestrei*; d. *Geastrum velutinum*; e. *Trichaster melanocephalus*; f. *Myriostoma coliforme*; g. *Bovista plumbea*; h. *Lycoperdon echinatum*.

Geastrum surtout, ainsi que le prouvent *G. hariotii* Lloyd (M.N.H.N.P., no. 310) (Planche VI, Fig. 3) et *G. caespitosum* Lloyd (M.N.H.N.P., no. 310) (Planche VI, Fig. 6) où se mêlent des ornements d'inégale hauteur.

Plus élevés, tous ces reliefs peuvent être assimilés à des piliers dont la variabilité est également considérable et la répartition systématique très vaste; ils sont en "borne" chez le *Tulostoma barlae* Quélet (M.N.H.N.P., Banyuls 1932) (Fig. 4b; Planche IV, Fig. 5), cylindriques sur les spores du *Geastrum pectinatum* Pers. (M.N.H.N.P., no. 312) (Planche VI, Fig. 4), subpiriformes chez le *G. saccatum* Fr. (R.H., L.M. 3086 C) (Planche VI, Fig. 2), en massue ou spatulés chez le *G. rufescens* Pers. (M.N.H.N.P., no. 312) et une espèce parfois séparée des Geasters sous le nom de *Trichaster melanocephalus* Czern. (M.N.H.N.P., no. 310; Mus. Hist. Nat. Vindobon. no. 3021) (Fig. 4e; Planche VII, Fig. 2). Par contre, les piliers qui hérissent la spore du *G. asper* (Mich.) Lloyd (M.N.H.N.P., Herb. Lloyd, no. 4398) (Planche VII, Fig. 1) offrent une assise plus évasée que l'extrémité obtuse ou presque tronquée. Une décoration analogue, quoique plus délicate existe chez les *Calvatia*: les petites épines, aiguës, mais larges et basses, à angle très ouvert, qui parsèment la spore du *C. candida* (Rostk.) Holl. (M.N.H.N.P., no. 320; Roumanie, no. 487), se retrouvent plus affirmées chez le *C. lilacina* (Berk. & Mont.) Henn. (M.N.H.N.P., no. 320). Les spores du rare *Queletia mirabilis* Fr. (M.N.H.N.P., no. 334; Montereau, Dumée, 8, 1913) (Planche IV, Figs 1, 2), portent également des épines fortes, hautes de 1–1·5 μ, étroitement coniques, parfois recourbées ou digitées; cette ornementation ne diffère pas fondamentalement de celle que révèle l'observation en PHM (Fig. 4a); cependant la structure fine des pics, comme zonée ou lamifiée, ne pouvait y être décelée. Une configuration sporale voisine apparaît chez le *Tulostoma floridanum* Lloyd (M.N.H.N.P., no. 332) (Planche IV, Fig. 3), appuyant ainsi les arguments d'ordre morphogénétique qui font entrer le genre *Queletia* dans les Tulostomataceae.

En fait, les ornements de forme simple sont rares et l'on observe plutôt leur fusion en masses plus complexes où se discerne encore chacune des unités primordiales. Ainsi, chez le *Queletia mirabilis*, les épines s'accolent ou se joignent par le sommet; chez le *T. floridanum*, elles se réunissent en faisceaux et constituent des crêtes verticalement striées (Planche IV, Figs 2, 3). Certaines spores de *Queletia* (Planche IV, Fig. 1) offrent même un relief tourmenté de massifs à arête crénelée. Ce phénomène est d'une grande généralité puisqu'il se retrouve dans le genre *Geastrum*; parmi les espèces étudiées, nous remarquons un arrangement des verrues, des tubercules et des piliers composés en massifs compacts, à bord festonné et surface mamelonnée, groupés en C ou en S, comme on le voit sur les spores du *G. velutinum* Morgan (M.N.H.N.P., no. 312) (Fig. 4d; Planche VI,

13. Étude Ornementale de Basidiospores au SEM

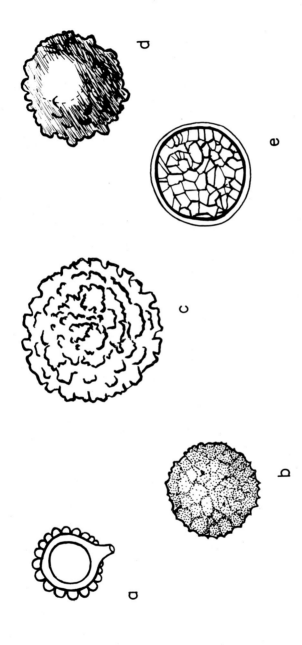

Fig. 5. La spore du *Mycenastrum corium* selon différents auteurs. D'après: a. Rolland (1906) ×1250; b. Coker et Couch (1928) ×1620; c. Gilbert (1950) ×2000; d. Pilát (1958) ×1500; e. Demoulin (1968) ×1500.

Fig. 5). Ces massifs sont à peine sinueux ou allongés sauf autour de l'appendice hilaire réduit à un bourrelet circulaire entourant le hile. La Planche VI (Figs 1 à 5) présente quelques aspects de la disposition étoilée, plus ou moins laciniée ou pétaloïde que revêt l'ornementation à partir de ce pôle proximal de la spore; ce dessin, nettement visible au SEM, demeure incertain au PHM (Fig. 4d). Si une indication de crêtes plus tortueuses et repliées se rencontre sur les spores du *G. pectinatum* (Planche VI, Fig. 4), c'est parmi les *Tulostoma* que nous trouvons la réalisation d'anastomoses entre les crêtes avec délimitation d'alvéoles, exceptionnels chez *T. chevalieri*, plus nombreux chez *T. barlae* et *T. berkeleyi* Lloyd (M.N.H.N.P., no. 332, Alabama; Type) où s'esquisse un réseau incomplet se ramifiant entre des ornements isolés. D'autres représentants du genre, tels *T. campestre* Morgan et *T. brumale* Pers. (M.N.H.N.P., no. 332) possèdent une ornementation sporale moins élevée, plus proche de celle du *T. chevalieri* (Planche IV, Fig. 4) ainsi qu'une couronne péri-hilaire. Nous avons également examiné au SEM les spores du *T. volvulatum* Borszcz. (M.N.H.N.P., no. 332) afin d'établir définitivement le caractère uni de leur surface. Subglobuleuses à piriformes ou cordiformes, elles sont parfaitement lisses ainsi que le PHM le laissait supposer.

C'est encore un réseau incomplet que le SEM a dévoilé pour les spores du *Calvatia lilacina* et des *Lycoperdon umbrinum*, *echinatum* et *atropurpureum* Vitt. (M.N.H.N.P., no. 314) considérées comme strictement verruqueuses d'après les observations à faible grossissement. En effet, la hauteur des connectifs qui relient

---

Planche V

Fig. 1. *Phellorina delestrei*, spores verruqueuses montrant un anneau péri-hilaire.
SEM, × *ca* 5000.
Fig. 2. *Ph. delestrei*, spore densément couverte de massifs verruqueux sillonnés.
SEM, × *ca* 10,000.
Fig. 3. *Ph. leptoderma*, ornementation de verrues arrondies et présence d'un appendice hilaire à peine marqué.
SEM, × *ca* 11,000.
Fig. 4. *Ph. leptoderma*, quelques aspects de l'ornementation sporale.
SEM, × *ca* 5000.
Fig. 5. *Whetstonia strobiliformis*, spores ornées de massifs verruqueux, parfois disposés en auréoles autour de l'appendice hilaire.
SEM, × *ca* 5000.
Fig. 6. *W. strobiliformis*, ornementation composée de grands massifs verruqueux séparés par des sillons à surface granuleuse.
SEM, × *ca* 10,000.

Clichés Laboratoire de Géologie du Muséum, Paris

13. *Étude Ornementale de Basidiospores au SEM* 273

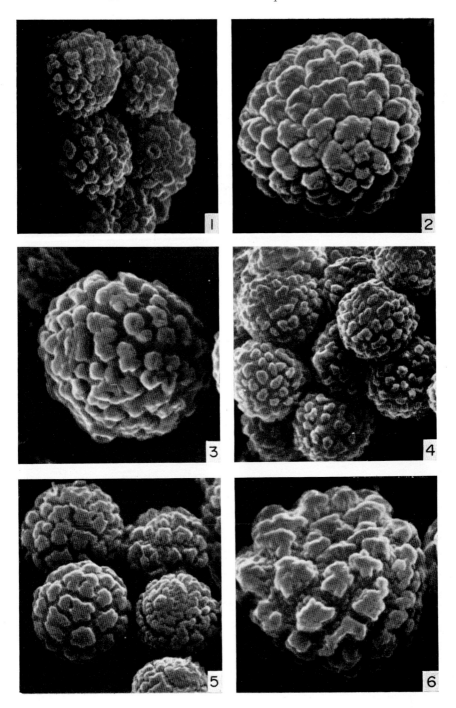

Planche V

les tubercules est souvent inférieure à 0·1 µ (Planche IX, Figs 1, 4, 5 et 6). La spore du *L. echinatum* se révèle donc ornée d'un réseau en chapelet interrompu dont les masses verruqueuses, toutes aplaties, mais diversement lobées, constituent les noeuds. Un schéma ornemental identique où cependant des épines et des masses triangulaires se sont substituées aux verrues se retrouve chez le *Lycoperdon atropurpureum*. Pour ces deux espèces, les fortes crêtes semblent creusées d'anfractuosités et parfois même évidées, aspect dont nous ne pouvons dire d'ailleurs s'il appartient à la réalité ou à un artefact.

Par contre, une telle configuration est bien réelle sur les spores du *Myriostoma coliforme* (Pers.) Corda (M.N.H.N.P., no. 308) dont l'ornementation remarquable s'ébauche peut-être chez le *Geastrum pectinatum* et surtout le *Trichaster melanocephalus* que certains auteurs tiennent comme intermédiaire entre *Geastrum* et *Myriostoma*. On y voit (Planche VII, Fig. 2) des piliers à méplat apical, s'évasant vers le sommet et le bas selon une concavité. Chez le *Myriostoma coliforme*, de telles colonnes délimitent des arcades dans de hautes ailes très importantes par rapport au diamètre du corps sporal. Epaisses, à sommet aplati, contournées, elles s'anastomosent irrégulièrement en alvéoles étroits mais profonds, s'isolant parfois en godets évasés, souvent doubles. Seule la photographie au SEM (Planche VII, Figs 3–6) pouvait mettre en évidence le lacis inextricable de ces crêtes ailées que l'observateur a peine à suivre au PHM en raison de la transparence de leur paroi (Fig. 4f).

Dans une étude magistrale qu'il consacra en 1935 à la famille des Phellorinés, Malençon décrivit comme réticulées leurs spores considérées auparavant comme

---

Planche VI

Fig. 1. *Geastrum subiculosum*, quelques spores ornées de verrues isolées, mais réunies en plaque étoilée autour de l'appendice hilaire.
SEM, × *ca* 10,000.

Fig. 2. *G. saccatum*, ornementation sporale composée de fortes verrues arrondies.
SEM, × *ca* 14,000.

Fig. 3. *G. hariotii*, piliers et granulations se réunissant en plaques festonnées.
SEM, × *ca* 10,000.

Fig. 4. *G. pectinatum*, ornementation hétérogène de piliers, crêtes et réseau incomplet.
SEM, × *ca* 10,000.

Fig. 5. *G. velutinum*, spores ornées de massifs verruqueux, disposés en étoile autour de l'appendice hilaire.
SEM, × *ca* 10,000.

Fig. 6. *G. caespitosum*, spore couverte d'ornements de tailles inégales.
SEM, × *ca* 15,700.

Clichés Laboratoire de Géologie du Muséum, Paris

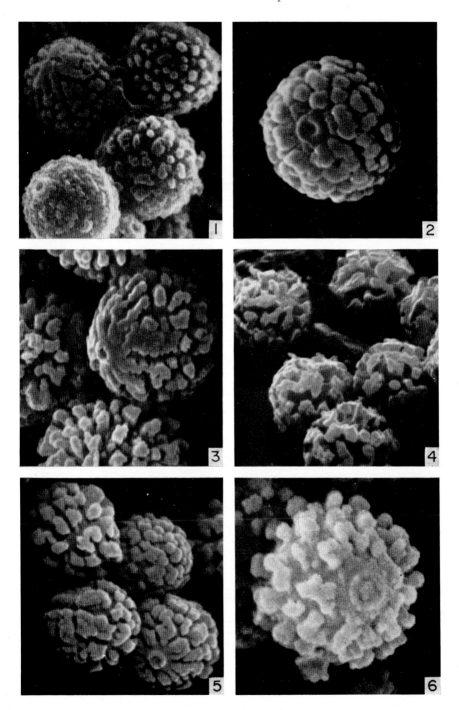

Planche VI

verruqueuses. Lorsque celles-ci sont montées dans un liquide pour l'observation au PHM la réfringence des téguments fait apparaître un réseau régulier à larges mailles légèrement anguleuses. Toutefois, après coloration avec le bleu coton lactique, ces mailles se teintent en bleu foncé et deviennent des plaques saillantes, polygonales-arrondies (Fig. 4c). Au SEM, nous avons examiné les spores sub-globuleuses, à peine étirées vers l'appendice hilaire à rebord annulaire, couvertes de massifs verruqueux, arrondis à anguleux, en V ou en Y (Planche V, Fig. 1) du *Phellorina delestrei* (Dur. & Mont.) Fischer (*Phellorina inquinans* Berk.) (M.N.H.N.P., no. 330; Algérie, L. Motelay; Soudan, M. Chudeau, 1909). Ces ornements se réunissent souvent en corolle autour de l'appendice hilaire et sont répartis de façon homogène à la surface de la spore si bien que les sillons qui les bordent, dessinent un réseau plus ou moins large. En effet, les spores sont parfois recouvertes d'un pavage serré de monticules festonnés-sillonnés (Planche V, Fig. 2). Chez le *Phellorina leptoderma* Pat. (M.N.H.N.P., no. 330; Djibouti 1893) (Planche V, Figs 3, 4), peut-être identique spécifiquement au précédent, la décoration est réalisée par des tubercules, plus espacés en général, plus arrondis, fréquemment agglomérés et faiblement déprimés en leur centre. L'ornementation en massifs et non en crêtes se reconnaît encore sur les spores du *Whetstonia strobiliformis* Lloyd (M.N.H.N.P., no. 330; Minnesota, Type) (Planche V, Figs 5, 6) qui est synonyme du *Dictyocephalus curvatus* Underw. (M.N.H.N.P., no. 330; Maroc, G. Malençon). Il s'agit d'épais et courts piliers, à peine bombés, à soubassement légèrement étréci, portant un plateau polygonal arrondi, plus ou moins échancré-festonné souvent; ainsi ornées, les spores globuleuses évoquent

---

Planche VII

Fig. 1. *Geastrum asper*, spore densément ornée d'épines.
SEM, × *ca* 9500.

Fig. 2. *Trichaster melanocephalus*, spore portant des piliers à sommet arrondi-élargi, se réunissant en crêtes.
SEM, × *ca* 11,000.

Fig. 3. *Myriostoma coliforme*, spore montrant une ornementation en réseau incomplet et irrégulier, avec formation de godets.
SEM, × *ca* 10,000.

Fig. 4. *M. coliforme*, spore avec des ailes perforées en arcades.
SEM, × *ca* 10,500.

Fig. 5. *M. coliforme*, autres aspects de l'ornementation.
SEM, × *ca* 5200.

Fig. 6. *M. coliforme*, détails de l'ornementation.
SEM, × *ca* 20,000.

Clichés Laboratoire de Géologie du Muséum, Paris

13. Étude Ornementale de Basidiospores au SEM

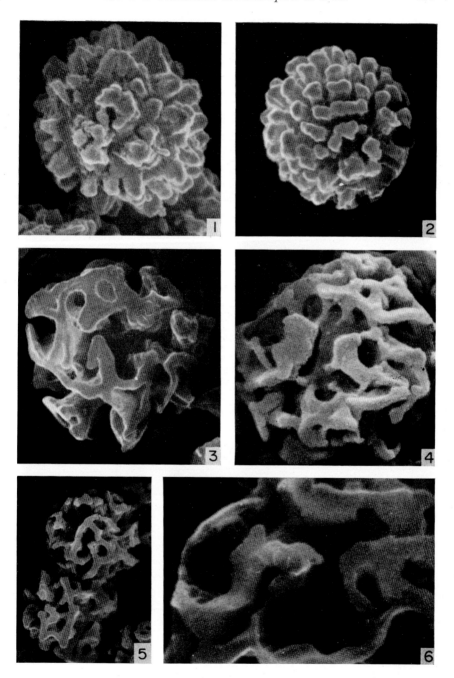

Planche VII

les cônes des Cupressacées. Parfois, l'ornementation apparaît plus contournée et dessine autour de l'appendice hilaire une double auréole ou une roue tandis que sur d'autres spores les sillons eux-mêmes se tapissent de nombreux tubercules perlés. Ainsi pour les trois représentants étudiés ici de la famille des Phellorinés le massif verruqueux est le motif fondamental de la décoration.

L'histoire de la description des spores chez le *Mycenastrum corium* (Guersent) Desvaux (M.N.H.N.P., no. 308; 1906, 1926) met particulièrement en évidence le rapide progrès apparu dans la connaissance de la topographie d'un matériel dès l'utilisation du microscope à balayage. Dans une période d'un peu plus d'un demi siècle, l'ornementation sporale de cette espèce, appartenant aux Lycoperdacées mais rappelant par certains caractères les Sclérodermes, a été représentée par différents auteurs ainsi que le montre la Fig. 5 à la p. 271. En 1906, Rolland dessinait, à petite échelle, le profil d'ornements en tubercules arrondis; plus tard, Coker et Couch (1928) distinguaient un réseau fin et compliqué alors qu'en 1950, Gilbert signalait la présence d'un réseau à mailles triangulaires juxtaposé à un relief verruqueux. Si Pilát (1958) n'indique que d'importantes verrues, sans plus de précision, Demoulin, en 1968, figure schématiquement un treillis complexe et anguleux. Au PHM, à fort grossissement, se discernent, sur les spores guttulées, globuleuses à ellipsoïdes, de 10–12 μ de diamètre, des alvéoles arrondis ou un réseau peu aisément déchiffrable et difficile à rendre par le dessin; le SEM seul (Planche VIII, Figs 1–4) pouvait apporter la confirmation de l'existence d'une nappe fovéolaire lisse ou à peine ondulée. Les fossettes, plus ou moins profondes,

---

Planche VIII

Fig. 1. *Mycenastrum corium*, détail de l'ornementation fovéolaire.
  SEM, × *ca* 10,000.
Fig. 2. *M. corium*, spore à surface granuleuse et alvéoles distants.
  SEM, × *ca* 5000.
Fig. 3. *M. corium*, ornementation fovéolaire.
  SEM, × *ca* 2400.
Fig. 4. *M. corium*, variations de l'ornementation fovéolaire.
  SEM, × *ca* 2500.
Fig. 5. *M. chilense*, ornementation fovéolaire et spinuleuse sur une même spore.
  SEM, × *ca* 4700.
Fig. 6. *M. chilense*, spore recouverte d'une nappe fovéolaire.
  SEM, × *ca* 5500.
Fig. 7. *M. phaeotrichum*, spores ornées de veines organisées en réseau.
  SEM, × *ca* 5000.

Clichés Laboratoire de Géologie du Muséum, Paris

13. *Étude Ornementale de Basidiospores au SEM* 279

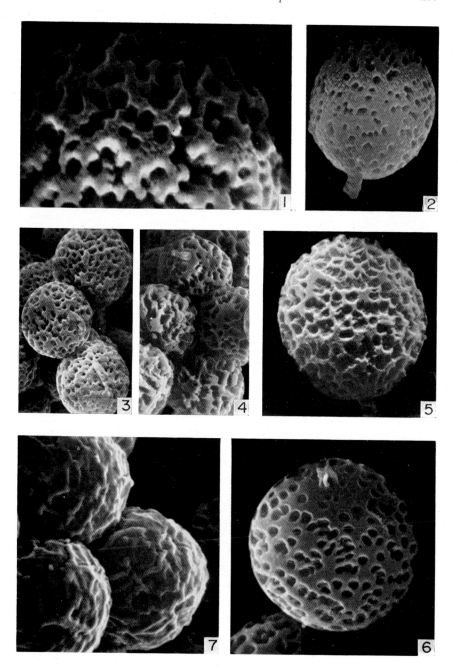

Planche VIII

de section circulaire, demeurent isolées ou bien se disposent en chaînes, fusionnent en anfractuosités lobées et parviennent ainsi à morceler la nappe ornementale en crêtes irrégulières, fait qui contribue à augmenter encore la variabilité du relief. Une spore (Planche VIII, Fig. 2), dont on observe nettement l'appendice hilaire cylindrique, montre une surface finement granuleuse autour des alvéoles comme il est possible de l'apercevoir, mais plus légèrement sur la photographie 1. Nous avons examiné les exemplaires du genre *Mycenastrum* que renferme l'Herbier du Laboratoire ; chez tous ceux qui appartiennent effectivement à ce genre, l'ornementation est de même type ; toutefois il semble bien, ainsi que plusieurs auteurs l'ont mentionné, que pour certains champignons, recueillis une seule fois en raison de leur rareté, souvent en mauvais état, il s'agisse du *Mycenastrum corium*. Les *M. spinulosum* Peck, *M. chilense* Montagne, *M. phaeotrichum* Berkeley (M.N.H.N.P., no. 308) ne seraient pas spécifiquement distincts. Chez le *M. chilense* (Planche VIII, Figs 5, 6), à grandes spores, la nappe fovéolaire demeure régulière ou est déchiquetée par la coalescence de très nombreuses fossettes ; sur quelques plages ne subsistent alors que des aiguillons très ténus, reliés par de fins tractus. Il faut noter que l'ornementation spinuleuse est donc là d'origine secondaire. C'est un réseau de veinules renflées aux noeuds qui recouvre les spores du *M. phaeotrichum* (Planche VIII, Fig. 7) ; les alvéoles y sont polygonaux-arrondis, parfois subdivisés et à peine creusés, rendant les

---

Planche IX

Fig. 1. *Calvatia lilacina*, ornementation de fortes épines et crêtes reliées par de fins connectifs.
SEM, × *ca* 10,500.

Fig. 2. *C. candida*, spores à épines arrondies isolées.
SEM, × *ca* 5700.

Fig. 3. *Bovista plumbea*, spores ornées de nombreuses petites verrues.
SEM, × *ca* 6200.

Fig. 4. *Lycoperdon umbrinum*, ornementation composée de tubercules reliés par des connectifs.
SEM, × *ca* 10,000.

Fig. 5. *L. echinatum*, spore ornée de tubercules aplatis, souvent coalescents et reliés par des tractus plus bas.
SEM, × *ca* 10,000.

Fig. 6. *L. echinatum*, quelques aspects de l'ornementation.
SEM, × *ca* 4800.

Clichés Laboratoire de Géologie du Muséum, Paris

13. *Étude Ornementale de Basidiospores au SEM* 281

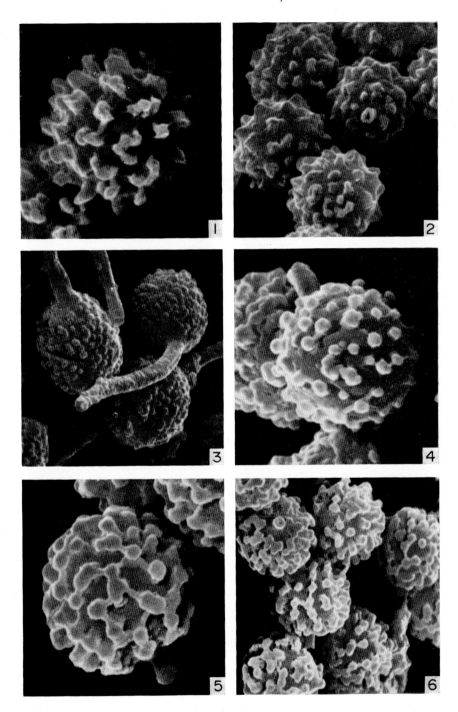

Planche IX

spores assez différentes de celles des précédentes espèces. Il semble d'ailleurs que la survivance d'une périspore comble et atténue les ressauts du relief.

Ce dernier exemple témoigne de nouveau, s'il en est besoin, de la valeur des observations au SEM et de l'amélioration qu'assure l'introduction de ce procédé. Approfondissement et découverte des détails, confirmation des corrélations entre les différents types ornementaux sont des résultats que nous avons obtenus lors d'investigations sur les spores des Basidiomycètes, en particulier, ici-même, des Russules, et auxquels souscrit entièrement cette étude consacrée aux spores de quelques Gastéromycètes. A la lumière des données nouvelles mises en évidence, l'interprétation des motifs ornementaux mérite d'être reconsidérée. De même qu'une schématisation ornementale succède aux nombreuses observations au PHM, il doit s'élaborer, au delà de l'analyse plus subtilement détaillée que livre le SEM, une synthèse architecturale plus précise pour chaque espèce. La valeur de l'ornementation sporale dans le domaine systématique en dépend.

TECHNIQUES UTILISÉES

Les spores étudiées proviennent de champignons desséchés ou d'exemplaires conservés dans un mélange alcool-formol-acide acétique, appartenant soit aux Herbiers du Laboratoire de Cryptogamie du Muséum National d'Histoire Naturelle de Paris (M.N.H.N.P.), soit aux collections personnelles de l'un de nous (R.H.). En microscopie photonique, les spores ont été observées, dessinées à main levée ou photographiées, alors qu'elles étaient montées dans le réactif de "Melzer" (Russules) ou le bleu coton lactique (Gastéromycètes). Pour les examens au microscope à balayage, les spores ont été prélevées sur les lamelles ou dans la gléba avec un pinceau humecté d'eau distillée, mais de préférence d'alcool à 70°, puis déposées sur une lamelle de verre, enfin soumises à dessiccation. Après ombrage, sous vide, à l'or-palladium, les observations ont été faites avec un microscope "Stereoscan" (Cambridge Instruments Company).

REMERCIEMENTS

Nous tenons à remercier tout particulièrement Mr. le Professeur Robert Laffitte, Directeur du Laboratoire de Géologie du Muséum pour l'accueil qu'il nous a réservé ainsi que pour les facilités de travail qu'il nous a offertes en nous permettant d'utiliser le SEM de son Laboratoire. Notre reconnaissance va également à Mlle D. Noël, à Mr. J. P. Bossy qui a la responsabilité, dans cette Chaire, des prises de vues au Stéréoscan et à Mme R. Haccard qui, au Laboratoire de Cryptogamie, a réalisé les tirages des photographies.

BIBLIOGRAPHIE

BIGELOW, H. E. et ROWLEY, J. R. (1968). Surface replicas of the spores of fleshy fungi. *Mycologia* **60**, 869–887.

COKER, W. C. et COUCH, J. N. (1928). "The Gasteromycetes of the Eastern United States and Canada", 202 pp. North Carolina University Press, Chapel Hill, N.C., U.S.A.

DEMOULIN, V. (1968). Gastéromycètes de Belgique: Sclérodermatales, Tulostomatales, Lycoperdales. *Bull. Jard. Bot. Nat. Belg.* **38**, 1–101.

DEMOULIN, V. (1969). Les Gastéromycètes. Introduction à l'étude des Gastéromycètes de Belgique. *Les Naturalistes Belges* **50**, 225–270.

DESVAUX, N. A. (1842). Sur le genre *Mycenastrum* du groupe des Lycoperdées. *Ann. Sci. Nat., Bot.*, ser. 2, **17**, 143–147.

DISSING, H. et LANGE, M. (1961). The genus *Geastrum* in Denmark. *Bot. Tidsskr.* **57**, 1–27.

DRING, D. M. (1964). Gasteromycetes of West Tropical Africa. *Myc. Papers*, **98**, 1–60.

ECHLIN, P. (1968). The use of the scanning reflection microscope in the study of plant and microbial material. *J. Microsc. Soc.* **88**, 407–418.

FISCHER, E. (1933). *In* Engler et Prantl, "Die natürlichen Pflanzenfamilien". (Ed. 2.)

GILBERT, E. J. (1950). *Mycenastrum corium* (Guersent) Desvaux. *Bull. Soc. Myc. Fr.* **66**, 101–105.

GREGORY, P. H. (1952). Fungus spores. *Trans. Br. mycol. Soc.* **35**, 1–18.

GREGORY, P. H. et NIXON, H. L. (1950). Electron micrographs of spores of some British Gasteromycetes. *Trans. Br. mycol. Soc.* **33**, 359–363.

HAWKER, L. E. (1968). Wall ornamentation of ascospores of species of *Elaphomyces* as shown by the scanning electron microscope. *Trans. Br. mycol. Soc.* **51**, 493–498.

HEIM, R. (1932). Mission Saharienne Augiéras-Draper 1927–1928. Champignons. *Bull. Mus.*, ser. 2, **40**, 915–932.

HEIM, R. (1938). "Les Lactario-Russulés du Domaine Oriental de Madagascar". Essai sur la classification et la phylogénie des Astérosporales, 196 pp. Lab. Crypt., Muséum, Paris.

HEIM, R. (1955). Les Lactaires d'Afrique Intertropicale (Congo Belge et Afrique Noire Française). *Bull. Jard. Bot. Etat. Brux.* **25**, 1–91.

HEIM, R. (1969). *Les Champignons d'Europe* (Ed. 2). 680 pp. N. Boubée et Cie, Paris.

HEIM, R. (1970). Particularités remarquables des Russules tropicales Pelliculariae liliputiennes: les complexes *annulata* et *radicans*. *Bull. Soc. Myc. Fr.* **86**, 59–77.

HOLLÓS, L. (1904). "Die Gasteromyceten Ungarns", 278 pp. Leipzig.

JONES, D. (1968a). Surface features of fungal spores as revealed in a scanning electron microscope. *Trans. Br. mycol. Soc.* **50**, 608–610.

JONES, D. (1968b). Examination of mycological specimens in the scanning electron microscope. *Trans. Br. mycol. Soc.* **50**, 690–691.

LLOYD, C. G. (1902). "The Geastrae." Mycological Writings, 44 pp. Cincinnati, Ohio, U.S.A.

LLOYD, C. G. (1906). "The Tylostomeae." Mycological Writings, 28 pp. Cincinnati, Ohio, U.S.A.

LONG, W. H. et PLUNKETT, O. A. (1940). Studies in the Gasteromycetes. I. The genus *Dictyocephalos*. *Mycologia* **32**, 696–709.

MALENÇON, G. (1935). Etude sur les Phellorinés. I. Le *Phellorinia Delestrei* (Dur. et Mtgn.) Ed. Fischer. *Ann. Crypt. Exot.* **8**, 5–48. II. Le *Dictyocephalus curvatus* Underwood. *Ibid.* **8**, 101–133.

MALENÇON, G. (1955). Le developpement du *Torrendia pulchella* Bres. et son importance morphogénétique. *Rev. Mycol.* **20**, 81–130.

PALMER, J. T. (1955). Observations on Gasteromycetes. 1–3. *Trans. Br. mycol. Soc.* **38**, 317–334.

PEGLER, D. N. et YOUNG, T. W. K. (1969). Ultrastructure of basidiospores in Agaricales in relation to Taxonomy and Spore Discharge. *Trans. Br. mycol. Soc.* **52**, 491–513.

PERREAU, J. (1967a). Observations sur la structure tégumentaire sporique des Astérosporales. *C.r. hebd. Séanc. Acad. Sci., Paris*, **260**, 4245–4248.

PERREAU, J. (1967b). Recherches sur la différenciation et la structure de la paroi sporale chez les Homobasidiomycètes à spores ornées. *Ann. Sci. Nat., Bot.* ser. 12, **8**, 639–746.

PERREAU, J. et HEIM, R. (1969). L'ornementation des basidiospores au microscope électronique à balayage. *Rev. Mycol.* **33** (1968), 329–340.

PILÁT, A. (1958). *Gasteromycetes*. Flora ČSR. 863 pp. Praha.

ROLLAND, L. (1906). Observations sur le *Mycenastrum corium* Desv. et sur le *Bovista plumbea* Pers. *Bull. Soc. Mycol. Fr.* **22**, 109–115.

ROWLEY, J. R. et FLYNN, J. J. (1966). Single stage carbon replicas of microspores. *Stain Techn.* **41**, 287–290.

UOZUMI, T. et FURUKAWA, H. (1969). Notes on the surface replicas of the spores of some species of higher Basidiomycetes. *Trans. Mycol. Soc. Japan* **9**, 145–150.

ZELLER, S. M. (1949). Keys to orders, families and genera of the Gasteromycetes. *Mycologia* **41**, 36–51.

14 | The Value of Scanning Electron Microscopy for the Examination of Actinomycetes

S. T. WILLIAMS and C. J. VELTKAMP

Hartley Botanical Laboratories, University of Liverpool, Liverpool, England

Abstract: Actinomycetes are related to bacteria but produce a range of sporing structures similar to those of fungi. The small size and delicate nature of these structures has made their observation in laboratory culture difficult and on natural substrates almost impossible. The scanning electron microscope (SEM) has proved to be a most useful tool for overcoming these difficulties.

Its efficiency for observation of actinomycetes is assessed and compared with other methods of light and electron microscopy which have been applied to these microorganisms. A number of examples of its application in studies of actinomycetes in both laboratory culture and on natural substrates is given.

INTRODUCTION

The actinomycetes, although long considered to be related to the fungi, are now regarded as an order, Actinomycetales, of the bacteria. They have hyphae with a prokaryotic internal structure and their cell wall composition is similar to that of Gram-positive bacteria. Nevertheless they do, like fungi, produce quite elaborate sporing structures, the range of which is still being increased as new forms are discovered. The morphology of the sporing structures has formed the basis of the classification of actinomycetes and despite the advent of new characters, distinction of genera is still largely made using morphological criteria (Baldacci and Locci, 1966; Williams *et al.*, 1968).

However, most actinomycete structures are considerably smaller ($0 \cdot 5$–$10 \cdot 0$ μm) and more delicate than those of fungi. This makes their examination more difficult. Observation of actinomycetes in laboratory culture with the light microscope can be unrewarding and has led in some instances to the erection of new genera on the basis of artefacts. Observation of these organisms growing on their natural substrates, for example in soil, is usually impossible without the introduction of artificial surfaces.

The scanning electron microscope can overcome many problems encountered

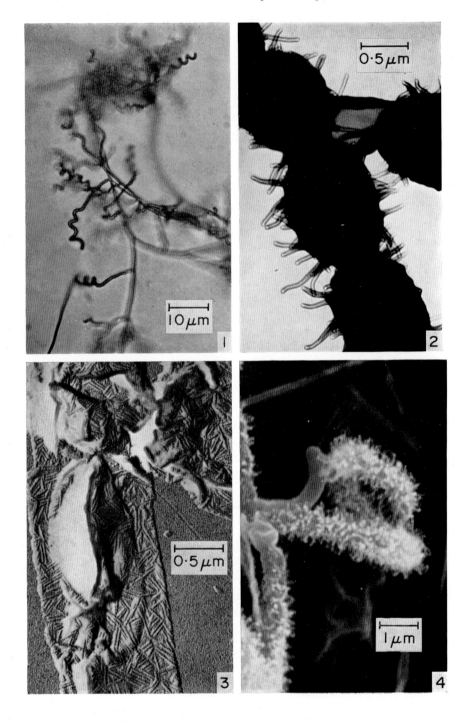

when examining actinomycetes with other techniques. It has, therefore, an important role to play in studies of the taxonomy, morphogenesis and ecology of these organisms.

COMPARISON OF TECHNIQUES FOR THE EXAMINATION OF ACTINOMYCETES

1. Light Microscopy

This is widely used for the routine examination of sporing structures and the identification of genera and species. Observations can be carried out in a number of ways. Cultures growing on an agar medium can be examined directly, using an objective lens of approximately ×40 magnification, thus obtaining information on intact sporing structures. Information is limited by lack of magnification and depth of focus (Plate I, Fig. 1). Similar observations can be carried out on slide or coverslip cultures, using methods such as those of Colmer and McCoy (1950) and Kawata and Shinobu (1959). If higher magnifications are required, preparations must be fixed and preferably stained before examination with an oil immersion objective. While this provides further information, disruption of sporing structures can occur.

Examination of growth on opaque natural substrates cannot usually be done using transmitted light and incident illumination is required. At the magnifications obtainable under such conditions, it is impossible to detect growth unless it is extremely dense.

2. Transmission Electron Microscopy

In addition to its important role in the examination of ultra-thin sections, the transmission electron microscope can also be used to observe intact structures. This may be done by direct examination of silhouettes of whole structures or by examination of replicas prepared from them.

Plate I

Fig. 1. Spore chains of a *Streptomyces* species growing on agar (light micrograph). ×900.

Fig. 2. Silhouette of *Streptomyces* spores showing surface spines (transmission electron micrograph). ×25,000.

Fig. 3. Carbon replica of *Streptomyces* spores showing rodlets (transmission electron micrograph). ×25,000.

Fig. 4. Spiral chain of hairy spores of *Streptomyces* species (scanning electron micrograph). ×10,000.

If spores of actinomycetes are picked up on electron microscope grids previously coated with a suitable film, such as carbon or Formvar, their silhouettes can be examined. This method has provided valuable information on spore surface ornamentation in the genus *Streptomyces* (Plate I, Fig. 2) (Küster, 1953; Preobrazhenskaya *et al.*, 1960; Tresner *et al.*, 1961). Usually gross features, such as shape of spore chains, are not observed and are determined by light microscopy. Additional information on a number of new genera has been provided by this method. Occasionally features observed in silhouette have been misinterpreted because of lack of information on the surface.

To obtain more information about the surface structure of electron dense material, it is necessary to prepare a replica of the surface with a film of carbon, using techniques similar to those described by Bradley (1960). The film can then be removed from the specimen and examined with the electron microscope. A few workers have used carbon replication to examine actinomycetes. Spores of *Streptomyces violaceoruber* were examined by Hopwood and Glauert (1961) and other species of this genus were examined by Dietz and Mathews (1962, 1968) and Preobrazhenskaya *et al.* (1965). This method is quite difficult and time-consuming, and if general information, such as spore shape or ornamentation, is all that is required, this can be obtained by other means. Dietz and Mathews (1969) found that scanning electron microscopy and carbon replication gave similar information on surface features of *Streptomyces* spores. Replication of certain structures, such as the fine hairs found on certain *Streptomyces* spores (Plate I, Fig. 4) is impossible. However, carbon replicas examined by transmission electron microscopy do reveal certain structures not seen by other methods. Spores of some *Streptomyces* species have been shown to have a pattern of small rodlets (approximately 10×100 nm) on their surface (Plate I, Fig. 3) (Dietz and Mathews, 1962; Bradley and Ritzi, 1968). These are not visible when whole spores are observed with the scanning electron microscope.

A further extension of replica methods is possible by use of freeze-etching techniques which have been successfully applied to a number of micro-organisms (Moor and Muhlethaler, 1963; Bauer, 1968; Ghosh *et al.*, 1969; Holt and Leadbetter, 1969). This technique, together with negative staining, has recently been used to study the surface structure of *Streptomyces* spores (Wildermuth, 1970).

None of the techniques considered in this section is applicable to actinomycetes growing on natural substrates.

3. *Scanning Electron Microscopy*

This was first applied to actinomycetes by Williams and Davies (1967) and its use has since been extended by Dietz and Mathews (1969), Williams *et al.* (1970)

and Williams (1970). It can provide information on gross features, such as spore chain shape, and at the same time finer details, such as spore ornamentation, can be observed (Plate I, Fig. 4). Thus it combines the roles of light microscopy and transmission electron microscopy of silhouettes, improving considerably on the accuracy of the information given. In addition, the same information is as easily obtained with actinomycetes growing on natural substrates as with laboratory cultures.

METHODS FOR THE PREPARATION AND EXAMINATION OF ACTINOMYCETES BY SCANNING ELECTRON MICROSCOPY

1. Growth of Cultures

To obtain the full advantages of scanning electron microscopy, it is essential that intact growth is examined. This can be achieved in a number of ways. Cover-slips of the same diameter as the specimen stubs can be inserted at an angle of 45° into agar medium and an actinomycete inoculated along the line where the upper glass surface meets the medium. Some growth occurs on the glass and remains on it when the cover-slip is carefully removed and fixed onto a specimen stub (Plate II, Figs 1 and 2).

In some cases it is possible to examine growth on agar media directly by removing a small cube of the medium (3–5 mm^3) with growth and fixing this to a stub (Plate II, Figs 3 and 4). This method is most satisfactory when growth occurring on the medium is not too dense.

Natural substrates can be fixed directly to specimen stubs or cut into suitable sized pieces.

2. Treatment of Specimens before Coating

In some instances, desiccation of specimens prior to coating results in considerable distortion, particularly to the vegetative hyphae. If distortion is unacceptable, it can be reduced or eliminated by treatment of the material before coating. The following procedure has proved useful.

Cover-slips or agar cubes with growth on them are fixed in a 1% solution of osmium tetroxide in buffer at room temperature for 2 hours. They are then washed 3 times with distilled water and dehydrated through a graded ethanol series (50% and 70% ethanol, 30 min in each, absolute ethanol, 60 min). The specimens in Plate II, Figs 1 and 3 were treated in this way. In a few cases changes induced by fixation have been observed. The fine hairs present on certain *Streptomyces* spores become brittle and break off (Williams and Sharples, 1970). It has been found that the sporing structures of some strains do not require any pre-treatment and with unexamined strains it is wise to observe both treated and untreated material.

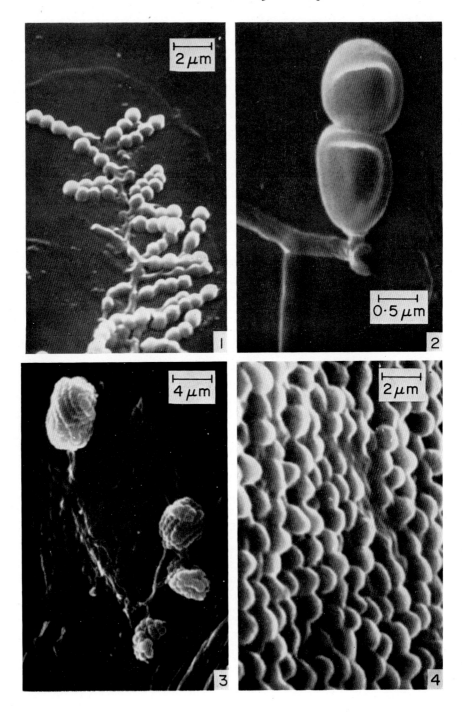

3. Operation of the Scanning Electron Microscope

Because of the small size of actinomycetes, magnifications needed for their examination are somewhat greater than those at which the instrument is often used. While it is possible to detect their growth en masse at magnifications between 500 and 1000 times, little detail of their structure is obtained below 5000. For finer details of surface structure, magnifications of 10,000 to 30,000 times are needed. For most studies, a working magnification of 10,000 times is required.

The instrument (Stereoscan, Cambridge Scientific Instruments Ltd.) is operated at 20 kV with a beam-specimen angle of 45° and a working distance of 9–11 mm. High condenser lens currents (0·7–0·9 A) are used to reduce the final probe diameter and increase resolution. Resulting noise levels on the visual screen are minimized by use of a large (200 μm) final aperture. Remaining noise is removed when photographing by choosing a long exposure time (200–400 sec).

APPLICATIONS OF SCANNING ELECTRON MICROSCOPY IN STUDIES OF ACTINOMYCETES

1. Routine Examination of Unknown Strains

Because of the speed and simplicity of the preparation procedures and the amount of accurate information readily obtained, the scanning electron microscope can be used for routine identification of actinomycetes. Using classification schemes based on the morphology of the sporing apparatus (e.g. Baldacci and Locci, 1966; Hütter, 1967; Williams et al., 1968), strains can be identified to the generic level (Plate II, Figs 1–4). In such schemes it is necessary to distinguish between spores developing on mycelium within the medium (substrate my-

---

Plate II

Fig. 1. Spore chains of *Micropolyspora* growing in cover-slip culture.
$\times 5000$.

Fig. 2. Paired spores of *Microbispora* growing in cover-slip culture.
$\times 20,000$.

Fig. 3. Sporangia of *Actinoplanes* growing from agar medium.
$\times 2500$.

Fig. 4. Spores of *Micromonospora* on the surface of agar medium.
$\times 10,000$.

celium) and those on mycelium above the medium (aerial mycelium). This can be most easily achieved by use of cover-slip cultures, the orientation of the cover-slip being noted as it is removed from the medium.

The rate at which such determinations can be carried out compares favourably with light microscopy, while at the same time providing increased information and accuracy. As large areas of the specimen can be examined, a more accurate idea of variation in morphological characteristics, such as spore ornamentation, within a culture can be obtained.

2. Provision of Additional Taxonomic Characters

As a result of investigations of various genera of actinomycetes, a number of new morphological characteristics has been observed. The shape of spores, previously seen at high magnifications only in silhouette, appears to be a more useful and stable characteristic than had been realized. Thus in the genus *Streptomyces*, spore shapes include spherical, square, oblong and irregular. Spores of two thermophilic genera, *Actinobifida* and *Thermoactinomyces*, were found to be ridged in a manner similar to certain *Bacillus* endospores (Williams, 1970). The characteristically paired spores of *Microbispora* were shown to be attached to their sporophores by an unusual ball and socket arrangement (Williams, 1970). With further study other features will no doubt be discovered.

3. Assessment of the Validity of Certain Taxonomic Characters

Due to the limitations imposed by light microscopy and examination of silhouettes by transmission electron microscopy, morphological features used to distinguish genera have sometimes been misinterpreted. The genus *Actinosporangium* was recognized by the formation of large "pseudo-sporangia" (Krasilnikov and Tsi-Shen, 1961). *Actinopycnidium* was distinguished by the formation of pycnidia containing spores (Krasilnikov, 1962). Observations of these genera by scanning electron microscopy indicated that in both, chains of spores were produced densely in localized areas; these intertwined giving the appearance of large structures (Williams, 1970). These organisms could be regarded as *Streptomyces* species, a suggestion which is supported by analysis of the composition of their cell walls (Becker et al., 1965; Yamaguchi, 1965).

The genus *Microechinospora* was distinguished from *Microellobosporia* by the occurrence of spines on its sporangia (Konev et al., 1965). Examination by scanning electron microscopy failed to detect spines and it could be regarded as a species of *Microellobosporia* (Williams, 1970).

For some years, certain *Streptomyces* species have been described as having "warty" spores from examination of their silhouettes. Observation of the surfaces of such spores by scanning electron microscopy indicated that some had a wrinkled but unornamented outer sheath, accounting for their irregular outline in silhouette.

4. *Studies of Spore Development*

Scanning and transmission electron microscopy can be profitably combined in studies of spore development. While the scanning electron microscope provides information on overall development and spore release, examination of sections by transmission electron microscopy gives details of internal changes.

As a result, a more accurate picture of the sequence of events can be obtained. Such a study was carried out on two *Streptomyces* species by Williams and Sharples (1970).

5. *Detection and Identification of Actinomycetes on Natural Substrates*

It is perhaps in studies of actinomycetes growing on their natural substrates that scanning electron microscopy has greatest potential. No other existing technique can provide as much information. Using the scanning electron microscope, micro-sites of growth, in some cases consisting of only one or two spore chains, can be detected on substrates such as soil particles, roots and fungal hyphae. Thus it is possible to study the behaviour of these micro-organisms in habitats commensurate with their size and sphere of influence. An actinomycete colony, once detected at relatively low magnification (Plate III, Fig. 1), can then be examined at progressively higher magnifications (Plate III, Figs 2, 3 and 4) to enable its identification to at least the generic level. The specimen illustrated is a *Streptomyces* species with spiral chains of smooth, angular spores, growing on a root fragment taken from soil.

The possible applications in studies of the ecology of actinomycetes are many, and valuable information has already been obtained in a number of investigations. Growth of acid-sensitive *Streptomyces* species in an acidic soil has been detected in micro-sites of ammonia accumulation on small pieces of organic matter (Mayfield, 1969). Selective transport of actinomycete spores through soil by arthropods was demonstrated by Ruddick (1969) and the same worker used scanning electron microscopy to show absorption of spores onto ion exchange resins and soil particles. At present, a study of the growth patterns of actinomycetes in natural and sterilized soil and live roots is in progress. Many other possible applications in ecological studies of actinomycetes and other micro-organisms exist and we can expect this technique to provide answers to many basic problems.

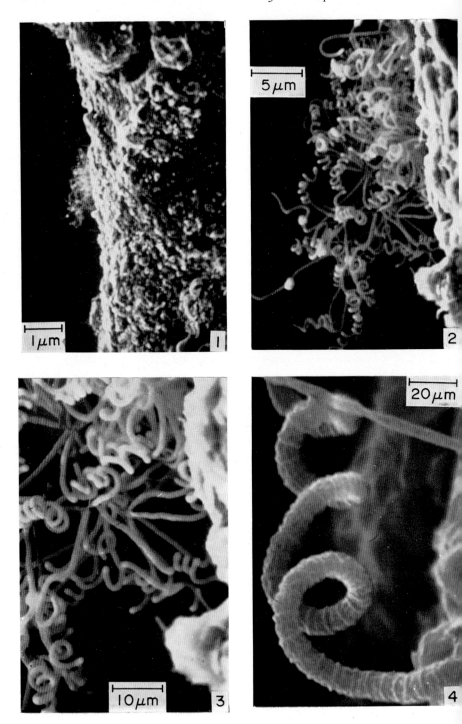

REFERENCES

BALDACCI, E. and LOCCI, R. (1966). A tentative arrangement of the genera in Actinomycetales. *Giorn. Microbiol.* **14**, 131–139.

BAUER, H. (1968). Interpretation of frozen-fractured membranes of *Lipomyces lipofer*. *J. Bact.* **96**, 853–854.

BECKER, B., LECHEVALIER, M. P. and LECHEVALIER, H. A. (1965). Chemical composition of cell-wall preparations from strains of various form-genera of aerobic actinomycetes. *Appl. Microbiol.* **13**, 236–243.

BRADLEY, D. E. (1960). IX. Replica techniques in applied electron microscopy. *J. r. microsc. Soc.* **79**, 101–118.

BRADLEY, S. G. and RITZI, D. (1968). Composition and ultrastructure of *Streptomyces venezuelae*. *J. Bact.* **95**, 2358–2364.

COLMER, A. A. and MCCOY, E. (1950). Some morphological and cultural studies on lake strains of *Micromonospora*. *Trans. Wis. Acad. Sci. Arts Lett.* **40**, 49–70.

DIETZ, A. and MATHEWS, J. (1962). Taxonomy by carbon replication. I. An examination of *Streptomyces hygroscopicus*. *Appl. Microbiol.* **10**, 258–263.

DIETZ, A. and MATHEWS, J. (1968). Taxonomy by carbon replication. II. Examination of eight additional cultures of *Streptomyces hygroscopicus*. *Appl. Microbiol.* **16**, 935–941.

DIETZ, A. and MATHEWS, J. (1969). Scanning electron microscopy of selected members of the *Streptomyces hygroscopicus* group. *Appl. Microbiol.* **18**, 694–696.

GHOSH, B. K., LAMPEN, J. O. and REMSEN, C. C. (1969). Periplasmic structure of frozen-etched and negatively stained cells of *Bacillus licheniformis* as correlated with penicillinase formation. *J. Bact.* **100**, 1002–1009.

HOLT, S. C. and LEADBETTER, E. R. (1969). Comparative ultrastructure of selected spore-forming bacteria: a freeze-etching study. *Bact. Rev.* **33**, 364–378.

HOPWOOD, D. A. and GLAUERT, A. M. (1961). Electron microscope observations on the surface structures of *Streptomyces violaceoruber*. *J. gen. Microbiol.* **26**, 325–330.

HÜTTER, R. (1967). "Systematik der Streptomyceten", 382 pp. Karger, Basel and New York.

KAWATO, M. and SHINOBU, R. (1959). On *Streptomyces herbaricolor* nov. sp. Supplement: A simple technique for the microscopical observation. *Mem. Osaka Univ. lib. Arts. Educ.* **8**, 114–119.

Plate III

Fig. 1. *Streptomyces* colony on dead root fragment taken from soil.
   × 650.

Fig. 2. *Streptomyces* colony on dead root fragment.
   × 1300.

Fig. 3. Spore chains within the colony.
   × 2600.

Fig. 4. Single spiral chain of smooth spores within the colony.
   × 10,000.

KONEV, Y. Y., TSYGANOV, V. A., MINBAEV, R. and MOROSOV, V. M. (1965). Isolation of a new genus of actinomycetes. IV *Scient. Congr. Leningrad Scient. Res. Inst. Antibiotics* 80–82.

KRASILNIKOV, N. A. and TSI-SHEN, J. (1961). *Actinosporangium* – a new genus of the *Actinoplanaceae* family. *Izv. Akad. Nauk. U.S.S.R. ser. Biol.* **1961**, 113–116.

KRASILNIKOV, N. A. (1962). A new genus of ray fungus – *Actinopycnidium* nov. gen. of the family *Actinomycetaceae*. *Mikrobiologiya* **31**, 249–253.

KÜSTER, E. (1963). Beitrag zur Genese und Morphologie der Streptomyceten-sporen. *VIth Int. Congr. Microbiol. (Rome)* 114–116.

MAYFIELD, C. I. (1969). A study of the behaviour of a successful soil streptomycete. Thesis. University of Liverpool.

MOOR, H. and MUHLETHALER, K. (1963). Fine structure in frozen-etched yeast cells. *J. Cell Biol.* **17**, 609–628.

PREOBRAZHENSKAYA, T. P., KUDRINA, E. S., MAKSIMOVA, T. S., SVESHINKOVA, M. A. and BOYARSKAYA, R. V. (1960). Electron microscopy of spores of various actinomycete species. *Mikrobiologiya* **29**, 51–55.

PREOBRAZHENSKAYA, T. P., MAKSIMOVA, T. S., LUKIVANOVICH, V. M. and EVKO, E. I. (1965). Application of the carbon replica method to the electron microscopic study of the surface of actinomycete spores. *Mikrobiologiya* **34**, 519–523.

RUDDICK, S. M. (1969). Some aspects of the structure and behaviour of actinomycete spores in soil. *Thesis, University of Liverpool.*

TRESNER, H. D., DAVIES, M. C. and BACKUS, E. J. (1961). Electron microscopy of *Streptomyces* spore morphology and its role in species differentiation. *J. Bact.* **81**, 70–80.

WILDERMUTH, H. (1970). Surface structure of streptomycete spores as revealed by negative staining and freeze-etching. *J. Bact.* **101**, 318–322.

WILLIAMS, S. T. (1970). Further investigations of actinomycetes by scanning electron microscopy. *J. gen. Microbiol.* **62**, 67–73.

WILLIAMS, S. T. and DAVIES, F. L. (1967). Use of a scanning electron microscope for the examination of actinomycetes. *J. gen. Microbiol.* **48**, 171–177.

WILLIAMS, S. T., DAVIES, F. L. and CROSS, T. (1968). Identification of genera of the Actinomycetales. *In* "Identification Methods for Microbiologists Part B" (Gibbs, B. M. and Shapton, D. A., eds), pp. 111–124. Academic Press, London and New York.

WILLIAMS, S. T., HATFIELD, H. L. and MAYFIELD, C. I. (1970). New methods for the observation of actinomycetes growing on natural and artificial substrates. *In:* "The Actinomycetales" (Prauser, H., ed.), pp. 379–391. Fischer, Jena.

WILLIAMS, S. T. and SHARPLES, G. P. (1970). A comparative study of spore formation in two *Streptomyces* species. *Microbios* **5**, 17–26.

YAMAGUCHI, T. (1965). Comparison of the cell-wall composition of morphologically distinct actinomycetes. *J. Bact.* **89**, 444–453.

15 | The Study of Fossil Epiphyllous Fungi by Scanning Electron Microscopy

K. L. ALVIN

Department of Botany, Imperial College of Science and Technology, London, England

Abstract: Fossil leaf fungi, except for certain Microthyriales from Tertiary angiosperm leaves, are poorly represented in the fossil record. Fossil Microthyriales often respond well to maceration techniques, and for this reason are among the best known of fossil fungi. Scanning electron microscopy has been used successfully to study some epiphyllous fungi on conifer shoots from the Wealden. These fungi do not respond well to maceration, and therefore give poor results with the light microscope. A preliminary study has been made of some undescribed Tertiary Microthyriales using both SEM as well as maceration and light microscopy. Results of the different methods are compared.

INTRODUCTION

A glance at the relevant pages of "The Fossil Record" (1967) is sufficient to indicate how woefully sparse is our knowledge of fossil fungi. Although there have been quite numerous reports of fungi in the tissues of vascular plants and on leaves, even in some instances in material as old as Devonian, comparatively few of these fungi are known in enough detail for them to be meaningfully compared with living fungi or classified with any precision.

The reasons for this are, firstly, that fungi are generally of soft structure lacking constituents highly resistant to decay such as the lignins and cutin of vascular plants, and therefore do not fossilize easily, and secondly, that since fungi are mostly microscopic, or practically so—only the fruiting structures of some of the higher fungi being truly macroscopic—fossil fungal remains cannot be adequately studied and are even unlikely to be encountered unless suitable microscopical methods are employed.

Nevertheless, Dilcher (1965), in his admirable summary of records of fossil leaf fungi, has shown that probably some twelve different orders of fungi are represented. These range from the chytrids through several orders of Ascomycetes to the Pucciniales and a number of different groups of Deuteromycetes.

It is therefore evident that the leaves of vascular plants have in the past, as today, provided suitable substrata or hosts for a wide range of different fungi and, moreover, that these fungi may, under certain conditions, fossilize sufficiently well to lend themselves to worthwhile study.

TRADITIONAL METHODS OF STUDY

Although some examples of fossil leaf fungi have been described from petrified material and studied in thin sections (e.g. Suzuki, 1910), they are more commonly found on leaf compressions, their tissues being preserved by carbonization in the same way as the tissues of the leaves on which they are situated. Such fossil material is usually completely opaque. It can be studied either by direct observation using reflected light or else by maceration and clearing techniques similar to those used in the preparation of fossil leaf cuticles for transmission light microscopy. The direct method of observation using reflected light is in practice limited by the difficulty of obtaining sufficiently high magnifications and enough depth of focus.

A large proportion of those fossil leaf fungi which are reasonably well known belongs to the ascomycete order Microthyriales, a group which is at present widespread chiefly on the leaves of angiosperms in the tropics. A peculiar property of many of these fungi is that the stromata seem to withstand oxidative and other maceration methods in much the same way as the cuticles and spores of vascular plants. Thus they have frequently come to light as a result of pollen and cuticle analysis of peat samples and similar material. Cookson (1947), Godwin and Andrew (1951) and others have described thyrothecia encountered in this way in isolation from the leaves on which they were borne and from any mycelium that they may have possessed. Fossil Microthyriales therefore may often lend themselves very well to the application of maceration methods on whole leaves or leaf fragments, and can sometimes be identified with considerable precision, especially if spores are preserved. Dilcher's (1965) work on Tertiary leaf fungi from Tennessee, in which some 20 different species are described (mostly Microthyriales) has been an outstanding contribution based on this approach.

RESULTS WITH THE SEM

My involvement with fossil epiphyllous fungi arose fortuitously during the study of a fossil conifer from the Wealden of the Isle of Wight. At an early stage in the investigation of this conifer, which was tentatively identified as a species of *Frenelopsis* Schenk, it had been noticed that many of the leafy shoots bore granules of carbonized substance up to 0·25 mm in diameter, and that these granules, which were usually scattered in patches, were associated with an obscuring of the host stomata within the patches. Using an ordinary stereoscopic microscope giving a magnification of up to about $\times 80$, it was impossible to resolve any details in the granules. When specimens were examined with the SEM, however, it soon became obvious that the granules were the fossilized fruit-bodies of a fungus, and, moreover, that they were connected with an epicuticular mycelium and associated with pycnidium-like structures. The mycelium was often thick, obscuring the host stomata completely. The fungus has now been formally described as *Stomiopeltites cretacea* Alvin and Muir (1970), and identified as a member of the sub-family Stomiopeltoideae of the Micropeltaceae (Microthyriales). It is the only representative of this order yet described from a deposit older than Tertiary.

Stomiopeltites cretacea was described almost entirely on the basis of observations made with the SEM. Plate I, Figs 1–4 show the kind of information revealed by the method. In Plate I, Fig. 1 shows a typical thyrothecium with a central ostiole and, on the left, an associated pycnidium-like structure. Higher magnifications (Plate I, Fig. 2) show the aggregated hyphae making up the thyrothecial wall. Septation of these hyphae cannot be seen, unless it is indicated by the very slight constrictions sometimes visible; that they form a pseudoparenchyma is shown by the cut specimen (Plate I, Fig. 4), and this interpretation is confirmed by light microscopical observations on macerated specimens. The connection of the thyrothecium with an epicuticular mycelium is clearly shown in Plate I, Fig. 3. Wherever well preserved thyrothecia are present, a mycelium is usually demonstrable on the host cuticle, and invariably pycnidium-like structures are also present.

Several attempts, using different maceration methods, were made to obtain useful preparations for light microscopy, but the fungus responded very poorly. The host cuticle is excellently preserved and easy to prepare by the standard method using Schultze's solution (nitric acid + potassium chlorate) followed, after a suitable time for oxidation of the carbonized material, by alkali. This completely destroys the fungus. Best results were obtained with plain nitric acid overnight followed by weak ammonia for a few minutes during which the specimens were watched under the microscope. As soon as the thyrothecia were clear enough to show some structure in transmitted light, the ammonia was washed away and the host cuticle mounted in glycerin-jelly. This method confirmed the pseudoparenchymatous structure of the thyrothecium, but yielded little or no further information; it completely destroyed the mycelium and eroded the edges of the thyrothecia. Clearly, the results obtained from the light microscopical methods alone would not have permitted us to describe the fungus in a worthwhile way.

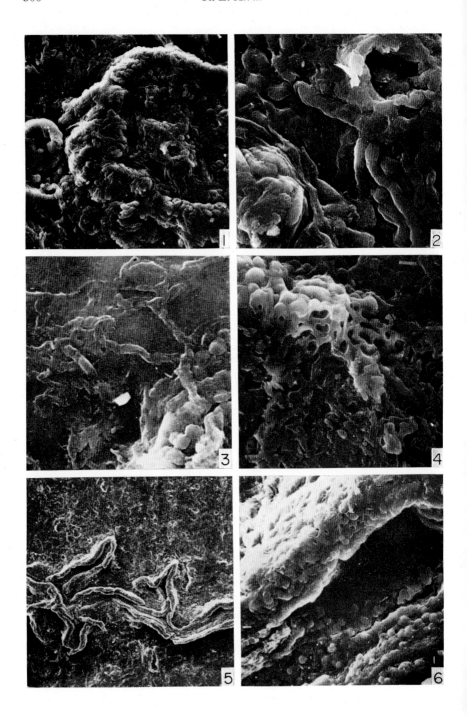

A fungus of a quite different kind has also been observed with the SEM on the same host shoot as *Stomiopeltites*. It is represented in Plate I, Figs 5 and 6. Unfortunately, I have so far found only three specimens, one of which was very poorly preserved, and no attempts have so far been made to prepare material for light microscopy. The fungus is tentatively identified as a member of the Hysteriaceae. The supposed ascocarps are elongated and branched, forming more or less stellate groups (Plate I, Fig. 5). Some evidence supporting this interpretation is obtained from higher magnifications (Plate I, Fig. 6) which show that the marginal "lips" (possibly the ascocarp wall) have a significantly smoother surface than the central region between the "lips". This trough (possible hymenial surface) is distinctly papillate, the papillae being of rather uniform size (about 5 μm in diameter). If the "wall" region is broken or cut it shows evidence of a pseudoparenchymatous structure, but no structure has yet been revealed in the "hymenium". The "ascocarps" are anchored to the host cuticle by hyphae which run out into an apparent epicuticular mycelium. The fruits are not erumpent through the host surface. This was made quite clear by macerating one specimen to dissolve away the fungus: no scar was left in the cuticle.

Some preliminary observations have also been made on some undescribed Microthyrialean fungi from a Tertiary deposit in Derbyshire. (The vascular

---

Plate I

Figs 1–6. Leaf fungi from the Lower Cretaceous (Wealden) of the Isle of Wight.

Fig. 1. *Stomiopeltites cretacea* Alvin and Muir. A thyrothecium and associated pycnidium-like structure. On the shoots of a conifer (cf. *Frenelopsis* Schenk).
× 360.

Fig. 2. *Stomiopeltites cretacea* Alvin and Muir. Part of thyrothecial surface showing aggregated hyphae forming the wall; top-right, the ostiole.
× 1450.

Fig. 3. *Stomiopeltites cretacea* Alvin and Muir. Part of host leaf surface with the epicuticular mycelium connecting with a thyrothecium (bottom-right).
× 1400.

Fig. 4. *Stomiopeltites cretacea* Alvin and Muir. Part of a cut thyrothecium showing the pseudoparenchymatous structure.
× 1400.

Fig. 5. Fungus possibly representing a member of the Hysteriaceae. On the same conifer.
× 54.

Fig. 6. The same fungus. Small part of the supposed hysterothecium showing, at the top, the presumed wall and, at the bottom, the possible hymenial surface.
× 430.

(Figs 1 and 2 are reproduced by permission of The Linnean Society of London.)

plant flora is currently under investigation by Boulter to whom I am indebted for the loan of material.) Unlike the Cretaceous fungi, these respond well to maceration, so results obtained with the SEM may be compared with those obtained by traditional methods.

One form, tentatively identified as a species of *Microthallites* Dilcher, was plentiful on a leaf resembling *Smilax*, and all stages in the development of the thyrothecium could be traced from apparent sporelings to old ones in which the central region had broken away leaving only a rim. Intermediate stages are represented in Plate II Figs 1–4 by both SEM and light photomicrographs. (Specimens for light microscopy were macerated in hydrogen peroxide.)

In Plate II, Figs 1 and 2 each show groups of fairly young thyrothecia, probably about half grown. The SEM shows that at this stage most of the surface, except for an area at the centre, is smooth. The central "lumpy" area is quite clearly delimited from the smooth peripheral part and seems to be relatively larger in the larger, presumably older, specimens. Thus, it is smallest in the specimen in the top left-hand corner, largest in the one at the top right, and somewhat intermediate in the one in the foreground in Plate II, Fig. 1. This differentiation does not obviously correspond to any feature observable in the light microscope (cf. Plate II, Fig. 3). Here there is a small group of cells at the centre with apparently thicker walls, but this area does not seem to increase in extent in larger specimens; on the contrary, it becomes a much less conspicuous

---

Plate II

Figs 1–7. Leaf fungi from the Tertiary (Lower Pliocene-Miocene) of Derbyshire.

Fig. 1. Cf. *Microthallites* Dilcher. Group of young thyrothecia, showing the surface differentiated into a smooth outer part and a tuberculate central area. On *Smilax*-like leaves.
×735.

Fig. 2. The same fungus. A thyrothecium probably representing a more mature stage.
×1410.

Figs 3 and 4. Light microscope photographs of specimens probably comparable with those in Figs 1 and 2 respectively.
Both ×500.

Fig. 5. Cf. *Callimothallus* Dilcher. Thyrothecium. On *Cryptomeria anglica* Boulter ms.
×435.

Fig. 6. The same fungus. Central portion of the thyrothecium showing possible pores.
×4150.

Fig. 7. Light microscope photograph of a thyrothecium of the same fungus.
×500.

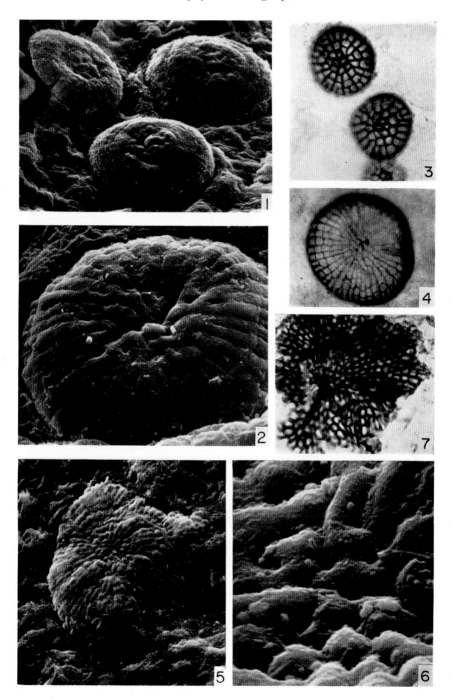

feature in those which are apparently much more mature (Plate II, Fig. 4).

At this more mature stage the SEM shows the entire surface elegantly sculptured in a regular radiating fashion, the furrows extending for varying distances from the periphery towards the centre. These furrows no doubt correspond to the radial walls of the files of cells seen in the light microscope (Plate II, Fig. 4). The tangential walls do not appear to be marked in the surface relief. This compares with the lack of any clear demarcation of the septa in the aggregated hyphae in *Stomiopeltites* (Plate I, Fig. 2). In the light microscope at this stage (Plate II, Fig. 4) most of the cells, except at the extreme periphery, are noticeably thinner-walled than earlier; this differentiation does not seem to be marked in the SEM view. In specimens where only a rim of peripheral cells remains, it is presumably the thin-walled cells which have collapsed and disappeared. Whether this represents a true final (dehiscence) stage is unknown.

Comparison between results with the SEM and the light microscope with this fungus raises some problems. The SEM view of the presumed younger specimens suggests that at an early stage the thyrothecium may be covered, except at the centre, with some kind of membranous envelope which becomes both thinner and proportionately narrower as the fruit grows and matures. The faint ring visible near the periphery in the specimen in Plate II, Fig. 2 may represent the remaining vestige of such a structure. No such investing membrane can be definitely seen in the light microscope preparations, although the smaller specimens (Plate II, Fig. 3) sometimes show a faint translucent "halo" not seen in the larger (Plate II, Fig. 4). It is possible that an investing membrane is being destroyed in maceration. The absence of any indication of tangential walls in the radiating dichotomizing components visible in the SEM of the mature fruitbody (Plate II, Fig. 2) contrasts markedly with the light microscope view (Plate II, Fig. 4) and also with the SEM view of the next fungus, in which the tangential walls are as clearly marked as the radial (Plate II, Fig. 5). The radiating "non-septate" system is comparable with that seen in the "basal layer" in Dilcher's species of *Microthallites*, i.e. the "layer" seen at a low focus in the light microscope. However, in the Derbyshire fungus, at no focus does one see a radiating system without tangential walls.

The fungus is tentatively classified on its general characters, including the absence of an epicuticular mycelium.

The other fungus examined from the Derbyshire deposit was found plentifully on leafy shoots of the conifer *Cryptomeria anglica* Boulter ms. Its classification is again only tentative, since investigations are not yet complete. The unmacerated fruit-body viewed with the SEM (Plate II, Fig. 5) shows typically a somewhat irregular shape. It appears distinctly domed, and the surface shows a clear pattern of radiating files of small protuberances, presumably the cells making up the pseudoparenchyma. There is obviously no central ostiole, but a close-up of the central region (Plate II, Fig. 6) shows a number of small pore-like depressions. These possible pores are highly reminiscent of the pores in Dilcher's *Callimothallus pertusus*. In the latter, the pores were present generally at the proximal ends of cells, especially in the central part of the stroma.

Macerated specimens seen in the light microscope (Plate II, Fig. 7) show again the irregular outline of the stroma and the radiating pattern of cells, of which there is apparently just a single layer. It is presumably the outer walls of these cells which are seen in relief in the SEM. The dome-shaped form of the stroma is less obvious, but can usually be demonstrated by careful focusing under high-power. The appearance of the cells themselves varies according to the method of maceration. In specimens macerated for a few hours in nitric acid followed by weak ammonia, a high proportion of the cells have perforated outer walls. (The specimen in Plate II, Fig. 7 was prepared in this way.) In those macerated in peroxide (2–3 weeks), only a very small proportion of the cells appear to be perforated—these chiefly towards the centre of the stroma. I believe therefore that nitric acid/alkali maceration causes damage to the cells, and that most or all of the perforations seen in specimens prepared by this method are artifacts; the frequently ruptured peripheral cells (Plate II, Fig. 7) supports this interpretation. Whether the few perforations seen in the peroxide macerated specimens are also artifacts and whether, if not, these are the same as the apparent pores in the un-macerated ones examined with the SEM are problems that remain to be solved.

CONCLUSIONS

The advantages of the SEM in the study of fossil epiphyllous fungi stem essentially from the fact that specimens can be examined directly without any maceration. The only prior treatment that may be necessary is demineralization in hydrofluoric acid to dissolve away particles of adhering inorganic matter. This however seems to have no significant effect on the preserved organic substances. Some fossil fungi, such as those investigated from the English Wealden, do not respond well to maceration techniques, but may reveal considerable detail of structure with the SEM. Comparison of results with the SEM with observations on comparable fungi made by traditional methods obviously raises problems, and it is clear that much work will have to be done on living forms using both methods, and also perhaps transmission electron microscopy, in order to interpret the results with the fossils meaningfully.

With material such as Tertiary Microthyriales which has traditionally been studied by maceration and light microscopy, the SEM may reveal structures not observable with the light microscope, such as the smooth peripheral zone in younger stromata of cf. *Microthallites*, but also, and no less importantly, the SEM should enable us to understand better the effects of maceration techniques, and to interpret more critically results obtained by these traditional methods.

REFERENCES

ALVIN, K. L. and MUIR, M. D. (1970). An epiphyllous fungus from the Lower Cretaceous. *Biol. J. Linn. Soc.* **2**, 55–59.

COOKSON, I. C. (1947). Fossil fungi from Tertiary deposits of the Southern Hemisphere. *Proc. Linn. Soc. N.S.W.* **72**, 207–214.

DILCHER, D. L. (1965). Epiphyllous fungi from Eocene deposits in western Tennessee, U.S.A. *Palaeontographica* (B) **116**, 1–54.

GODWIN, H. and ANDREW, R. (1951). A fungal fruit body common in postglacial peat deposits. *New Phytol.* **50**, 179–183.

HARLAND, W. B. *et al.* (eds) (1967). "The Fossil Record." Geological Society, London.

SUZUKI, Y. (1910). On the structure and affinities of two new conifers and a new fungus from the Upper Cretaceous of Hokkaido (Yezo). *Bot. Mag. Tokyo* **24**, 181–196.

16 | Preparation of Labile Biological Material for Examination in the Scanning Electron Microscope

PATRICK ECHLIN

Department of Botany, University of Cambridge, Cambridge, England

INTRODUCTION

One of the main disadvantages to the examination of biological material at high resolution in both the transmission and scanning electron microscope is that the observations usually have to be made on material in a highly dehydrated state. Water is an important structural component of most biological material, and elaborate steps are taken to ensure that it is removed from plant and animal material prior to their examination in the high vacuum of electron probe instruments. Ideally, this dehydration should be accomplished without causing any specimen damage.

Certain biological materials, such as wood, the hard outer covering of seeds and pollen grains, diatoms and bone are readily examined in the scanning electron microscope. The very fact that they are hard means that they are able to resist collapse following the removal of the "structural water". Fortunately these tough resistant structures frequently contain the morphological characters which are of vital importance in taxonomy and systematics, and already we have a considerable corpus of knowledge gleaned from such features. Plant cells are generally easier to examine in the scanning microscope simply because they possess a rigid cellulosic cell wall, which may be marked or impregnated and encrusted with such resistant material as lignin, suberin and sporopollenin. Difficulties arise when hydrophytes are examined, for although they have cellulosic cell walls, their internal water is of vital structural importance. With the exception of hard animal tissue, such as teeth and bone, which are strengthened with calcium salts, most animal cells are extremely prone to collapse in a high vacuum. This same sensitivity is shared by many microbes and protozoa.

It is thus usually necessary to pre-treat delicate biological material before it is examined in the scanning microscope. Unless adequate care is taken to preserve the three dimensional topography of delicate specimens, the taxonomic use of any distinctive characters will be obliterated by the artefact of collapsed cells. The adequate preservation of the surface is vital to the meaningful interpretation of any image obtained by electron probe instrumentation. Yet it is not always necessary to preserve the surface of the entire structure, because, depending on the level at which detail is being sought, the collapse of the whole cell or tissue may not be manifest. If, however, the interrelationship of prominent features over a wide area is important, then although this large area must be adequately preserved, the intervening details are of little consequence. Although it may be dangerous to make "general rules" for any biologic preparative technique, it is probably only necessary to ensure that the preparative technique maintains preservation to about $\frac{1}{5}$ of the size of the features being examined. Thus, there seems to be little point in ensuring faithful preservation of material down to the 10 Å level when the feature to be examined is of the order of 100 nm.

A lot has already been written on the various preparative techniques, and it is impossible to cover all the methods which have ever been used. For those readers who are interested in a technique as applied to specific plant, animal or microbial material it is suggested that they refer to the excellent bibliographies compiled by Rossi (1968), Johnson (1969) and Wells (1969, 1970). More general references to methods which have been applied to larger groups of organisms are given by Echlin (1968), Boyde and Barber (1969), Boyde and Wood (1969), Heywood (1969), Marszalek and Small (1969) and Small and Marszalek (1969). The paper by Boyde and Wood (1969) is particularly useful, as it contains an excellent review of the techniques for preserving soft animal tissue, most of which have been pioneered and tested by Boyde and his co-workers.

The purpose of this present paper can be no more than to describe in outline the general type of methods which have been utilized in the preservation of labile biological material, and this will be followed by a brief discussion of some newer techniques which are becoming available. The methodology which follows is only really applicable to the examination of material which is to be examined in the emissive mode of operation. When the cathodeluminescent and X-ray microanalysis modes are to be used, it may be necessary to modify these methods.

MINIMAL PREPARATION

Before any preparative technique is initiated, it is necessary to ensure that the surface of the specimen is free of any extraneous and foreign material. With dry resistant specimens this may be accomplished by a jet of clean air across the specimens once they have been firmly affixed to the specimen stub. The small cans of ultra clean Freon gas which are commercially available are ideally suited for this purpose. With wet specimens it is necessary to wash the sample in a suitable ionically balanced solution, which is then followed by a brief rinse in distilled water. The latter treatment will remove any salt crystals which may obscure surface detail. For strongly adherent material it may be necessary to carry out a more turbulent washing procedure, or even resort to a brief exposure to ultrasonic waves.

As previously indicated, most plant material can be examined with no more treatment than controlled air drying followed by the use of chemical desiccants. Even some aquatic plants have been examined by these methods, and Heslop-Harrison and Heslop-Harrison (1969) were able to observe considerable detail in an untreated leaf of the xerophyte *Dianthus plumarius* and attributed some of their success to the ability of the vacuum system in the Stereoscan scanning electron microscope to take up any water which is evaporating from the specimen. With mesophytes and hydrophytes, such as *Elodea*, although there was considerable cellular collapse, it was still possible to make out the general configuration of the tissue. The presence of dissolved minerals and ions in the cell sap may well have contributed to the generation of secondary electrons as well as providing a suitable conductive pathway.

FIXATION AND DEHYDRATION TECHNIQUES

A measure of success has been obtained using some of the preparative methods devised for the examination of material at the level of the light microscope. Such procedures are adequately described in books on preparative microscopy and will not be dealt with here. Following the fixation, it is necessary to dehydrate the specimens, and this is accomplished by passage through graded ethanol or acetone. Considerable success has been obtained from an examination of wax-embedded material. The samples are prepared in the usual way, embedded in wax and thin sections cut using a glass knife to lessen the chance of knife scratches. The sections are then dewaxed and may be examined in the scanning microscope in the usual way. This procedure is particularly useful if one wishes to examine the internal detail at medium resolution. Better results have been obtained using the preparative techniques which have been applied to material prior to its examination in the transmission electron microscope. There are many recipes available, but the general concensus is that a fixative containing an organic aldehyde such as formaldehyde or glutaraldehyde probably gives the best results. The organic aldehydes tend to toughen tissue and this lessens the degree of collapse which usually occurs in the ensuing dehydration procedures. A number of workers recommend the use of hardening agents such as Parduczs fixatives which tend to harden biological tissue. It is clear, however, that the dehydration procedure is where the damage occurs in spite of the seemingly judicious application of fixatives. The answer would seem to lie in adopting dehydration techniques which do not involve distortion of the material by surface tension forces which occur as the receding liquid evaporates.

FREEZE DRYING

In this procedure the sample, which may or may not have been prefixed, is rapidly frozen, and whilst still frozen, the water removed by sublimation in a high vacuum. In order to minimize ice crystal damage it is necessary to lower the nucleation rate of water, and this may achieved by a number of methods. It is possible to infiltrate the tissue with physiological amounts of cryoprotective substances such as glycerol, dimethyl sulphoxide or polyvinylpyrrolidone, all of which substantially lower the rate of ice crystal formation. Unfortunately, the low vapour pressure of the chemical precludes their presence in samples which are going to be introduced into the microscope column operated at room temperatures because they are a source of contamination and may seriously obscure specimen detail. Boyde and Wood (1969) claim to have achieved a lower nucleation rate by using a 2% chloroform solution, the advantage being that the chloroform would be more easily removed in the high vacuum. Moor (1970) has demonstrated that it is possible to lower the critical freezing rate other than by using cryoprotective substances. Such methods include ultra rapid quench freezing at a rate of at least 10^4 °C/sec and the application of high pressure during the freezing process. It is now possible to obtain water vitrification to a depth of about 200 micrometres from the surface which will cover most biological applications. Boyde and Wood (1969) have described in detail the procedures involved in freeze drying, especially as applied to the preparation of material for examination in the scanning electron microscope, and reference should be made to this paper.

CRITICAL POINT DRYING

Horridge and Tamm (1969) have applied the technique originally devised by Anderson (1951) to the preparation of the protozoan *Opalina*. The organisms were fixed and then dehydrated in alcohol. The alcohol was replaced by some non-polar solvent such as amyl acetate and then the amyl acetate and organism placed in a critical point apparatus. The bomb was flushed out with liquid carbon dioxide at room temperature to replace the amyl acetate, and once the amyl acetate had been removed the bomb was sealed and the temperature raised to 50° C. During the heating the carbon dioxide goes through its critical point and forms a gas without the formation of a phase boundary. The pressure was then released, the CO_2 allowed to escape and the dried organisms were placed on the specimen stub. The preparative procedure preserves the fine structural detail as well as the metachronal waves of the ciliary action. Boyde and Wood (1969) have also used this procedure and while they attest to its usefulness, point out that if not carefully applied may easily give rise to artefacts and result in bulk shrinkage. At present, too little work has been carried out using the critical point drying method to come to any firm conclusion regarding its usefulness. The principal advantage of the procedure would appear to be that it is easy to assemble the necessary apparatus and it is a relatively quick procedure.

All the methods so far described suffer from the disadvantage in that the tissue being examined must be sufficiently robust to retain its natural configuration in the absence of structural water. This appears to be the case with plant cells which, as has been mentioned previously, are constrained by a cellulosic wall. Many animal cells may also retain their natural tridimensional configuration, but for those that do not it is necessary to allow the water to remain inside. To examine such wet specimens it is necessary to use specially designed wet chambers or to cool the specimen to a point where the vapour pressure of water is so low that it defies rapid sublimation in the relatively high vacuum in the scanning electron microscope.

EXAMINATION OF LIVING MATERIAL

The scanning electron microscope has several advantages over transmission electron microscopy as far as examining living specimens. Firstly, it is not necessary to cut sections as the specimen may be examined in its entirety, and secondly, the beam current is several orders of magnitude lower than in the transmission microscope. Concomitant to this there is a considerable reduction in the amount of energy impinging on the specimen. The only organisms which have successfully been examined in the scanning microscope are specimens which can either retain their water against a high vacuum, or can undergo a considerable dehydration/hydration cycle and still remain viable. Thus Hayes and Pease (1968) were able to demonstrate that all stages of development of *Tribolium confusum* were able to survive within a vacuum of 10^{-3} torr, and Humphreys, Hayes and Hughes (1967) were able to examine untreated *Artemia* cysts in the scanning microscope and demonstrated that the organisms resumed activity following hydration. For the examination of wet and labile material it is necessary to use a wet stage in which the organisms remain in an environment of relative high humidity. Although several prototype stages have been suggested and some have been constructed, there is presently insufficient experimental evidence to warrant further discussion of this method of specimen examination.

EXAMINATION OF MATERIALS AT LOW TEMPERATURES

In collaboration with physicists from Cambridge Scientific Instruments, some preliminary studies have been carried out on an examination of material at low temperatures inside the microscope column. Full details of the instrumentation and technique are given by Echlin *et al.* (1970), and only the preparative procedures will be discussed here. The specimens were prepared by either of the following two methods.

Initially it was found that biological material could be quench-frozen by plunging specimen stubs with the material mounted on the surface into liquid nitrogen at $-196°$ C. With plant material such as leaves of *Lagarosiphon major* and *Tradescantia bracteata* and root hairs of *Sinapis alba* and *Zea mays*, adequate preservation was obtained with little or no evidence of ice crystal damage. It is clear that the cellulose wall of plant cells is sufficiently robust to prevent or retard extensive ice crystal formation and growth.

These procedures were unsatisfactory for optimal preservation of microalgae, protozoa and animal tissue and cells. Such extremely hydrated and delicate tissues were drained or damp-dried and either placed directly onto cooled aluminium specimen stubs or onto clean platinum discs which in turn were fixed to cooled specimen stubs by a spot of silver-dag. The specimen stubs were then rapidly plunged into liquid Freon 22 (monochlorodifluoromethane) maintained at its melting point of $-140°$ C by liquid nitrogen. The Freon 22 quench freezing invariably gave specimens free of ice crystal damage. A mixture of isopentane and methylcyclohexane although an adequate quench coolant, was unfortunately transferred via the specimen stub into the microscope and it was impossible to maintain an adequate vacuum in the instrument.

Following the quench freezing the specimen stubs were quickly (within 4–5 seconds) transferred into the microscope, maintained at $-180°$ C with liquid nitrogen and the column pumped down to 1×10^{-4} torr its working vacuum. An immediate examination of the specimens revealed that they were covered with ice. This water is derived from two sources. In the case of aquatic organisms or cells maintained in an aqueous medium, the water was part of their immediate environment. Such organisms were either examined in their natural state, or carefully washed in isotonic salts solution to remove any surface debris. A certain amount of atmospheric water will condense onto the specimen during the transfer from the quench coolant to the specimen stage. We were initially concerned at the presence of this water, and elaborate steps were taken to construct a dry environment specimen transfer chamber. But as will be shown it is relatively easy to remove the surface water within the microscope so these cumbersome procedures were abandoned. The temperature was raised to between $-100°$ C and $-90°$ C whereupon the high vacuum within the microscope column caused the surface water to sublime revealing the tissue surface below. The removal of water had to be carefully monitored, and the temperature was not allowed to rise above $-85°$ C. If too much water was removed, by allowing too great a rise in temperature, then the tissue collapsed. The same tissue showed considerable collapse and surface artefacts when fixed in buffered glutaraldehyde and then taken through a freeze-substitution procedure. Examination of material at low temperatures dramatically demonstrates that water is a vital structural component in many plant and animal cells.

Surprisingly large amounts of water could be removed from the specimen and in no instance did we observe or experience inoperatively high pressures. When the removal of the surface water was completed, and this usually took place within 3–5 minutes of placing the specimen into the microscope, the temperature was maintained at between $-140°$ C and $-100°$ C during the examination of the specimen.

All the specimens were examined in the emissive mode of operation over a wide range of accelerating voltages without the benefit of surface coating. Boyde and Wood (1969) and Echlin (1970) have discussed in some detail the advantages of examining biological specimens at low accelerating voltages. Attempts were made to spray "Duron" antistatic spray onto the surface of the *Tradescantia* leaves prior to quench freezing. This procedure was abandoned as it gave such consistently poor results because the micro-droplets of the antistatic agent obscured any surface detail. In spite of the absence of a surface coating it has been possible to obtain a resolution of the order of 1000 Å. Although accelerating voltages as high as 30 kV were employed, it was usual to take pictures at 20 kV or lower as this reduces the energy impinging on the specimen. It was possible to measure the radiation dosage per unit area falling on the specimen—a feature of particular importance when examining biological material.

This technique has been particularly useful in examining microalgae and protozoa, because they retained their three-dimensional shape while in the frozen state. It was possible to see details and the orientation of the pellicular warts on the myonemes of *Euglena* and the array of cilia on *Paramecium*. Preparation of these objects by more conventional means invariably resulted in serious collapse.

The technique has been particularly useful in examining tissue culture cells. Mouse L cells, which are sensitive to adverse conditions, were grown on platinum discs, washed several times with a buffered isotonic salts solution, before being quench frozen. The cells retain their natural configuration, and show such features as the nuclear bulge and the delicate protoplasmic protuberances which are believed to be associated with the attachment of the cells to the substrate. It is also possible to see the fine filamentary appendages at the tips of the cells. Examination of the sample at different tilt angles relative to the primary electron beam revealed further information about the specimen. These preliminary studies clearly demonstrate the usefulness of the technique in the examination of biological material and the instrumentation and procedures employed ensure that the specimens faithfully retain their natural three-dimensional configuration. The only treatment which has been applied to the specimens is a quench freezing technique which has already proven satisfactory in the preservation of cells and tissues. The results so far obtained are often better than results obtained by fixation and freeze drying techniques. An advantage of the latter technique is that specimens may be coated before examination, thus improving both resolution and ease of interpretation. A distinct advantage is that images may be obtained within a few minutes of the specimen being removed from its natural environment.

SURFACE COATING

Following the adequate preparation of the specimens by which ever means is considered best, it is necessary to coat the specimen with a layer of suitable conducting material prior to its examination in the microscope. This layer, which is usually in the form of evaporated carbon and/or a heavy metal such as gold, provides a route for the transference of excess thermal and electrical build up, both of which may considerably distort the final image. A number of surface coatings and the techniques for their application are now available, and for further detail reference should be made to the publications at the beginning of this present paper. In some instances it is not practicable to apply a surface coating, and in order to avoid the adverse charging effects it is necessary to operate at considerably reduced accelerating voltages. We have carried out a series of experiments in which fresh material was examined uncoated at an accelerating voltage of between 1 and 2 kV at a resolution of between 80–90 nm. There is considerably less beam penetration and, more importantly on biological material, a considerable reduction in the radiation flux impinging on the specimen.

The advantage of the low voltage operation is that it allows a rapid examination of material at medium resolution without the disadvantage of the time consuming preparative and coating techniques. Even fairly labile specimens may be examined provided this is done within a few minutes of their being introduced into the microscope column.

REFERENCES

ANDERSON, T. F. (1951). Techniques for the preservation of three-dimensional structure in preparing specimens for the electron microscope. *Trans. N.Y. Acad. Sci.* Ser. 11, **13**, 130.

BOYDE, A. and BARBER, V. C. (1969). Freeze drying methods for the scanning electron microscopical study of the protozoan *Spirostomum ambiguum* and the statocyst of the cephalopod mollusc *Loligo vulgaris. J. Cell. Sci.* **4**, 223–239.

BOYDE, A. and WOOD, C. (1969). Preparation of animal tissues for surface scanning electron microscopy. *J. Microsc.* **90**, 221–249.

ECHLIN, P. (1968). The use of the scanning reflection electron microscope in the study of plant and microbial material. *J. r. Microsc. Soc.* **88**, 407–418.

ECHLIN, P. (1970). The application of scanning electron microscopy to biological research. *Proc. r. Soc. Lond.* (In press.)

ECHLIN, P., PADEN, D., DRONZEK, B. and WAYTE, R. (1970). Scanning electron microscopy of labile biological material maintained under controlled conditions. *Proceedings 3rd Annual Scanning Electron Microscopy Symposium*, pp. 49–56. IIT Research Institute, Chicago.

HAYES, T. L. and PEASE, R. F. W. (1968). The scanning electron microscope: principles and applications in biology and medicine. *Adv. Biol. Med. Phys.* **12**, 85–137.

HESLOP-HARRISON, Y. and HESLOP-HARRISON, J. (1969). Scanning electron microscopy of leaf surfaces. *Proceedings 2nd Annual Scanning Electron Microscopy Symposium.* pp. 119–126, IIT Research Institute, Chicago.

HEYWOOD, V. H. (1969). Scanning electron microscopy in the study of plant materials. *Micron* **1**, 1–14.

HORRIDGE, G. A. and TAMM, S. L. (1969). Critical point drying for scanning electron microscopic study of ciliary motion. *Science, N.Y.* **163**, 817–818.

HUMPHREYS, W. J., HAYES, T. L. and PEASE, R. F. W. (1967). *Proceedings 25th EMSA Meeting Chicago.* p. 50. (C. Arceneaux, ed.).

JOHNSON, V. (1969). Bibliography on the scanning electron microscope. *Proceedings 2nd Annual Scanning Electron Microscopy Symposium.* pp. 483–525, IIT Research Institute, Chicago.

MARSZALEK, D. S. and SMALL, E. B. (1969). Preparation of soft biological materials for scanning electron microscopy. *Proceedings 2nd Annual Scanning Electron Microscopy Symposium.* pp. 231–240, IIT Research Institute, Chicago.

MOOR, H. (1970). Recent progress in freeze etching technique. *Proc. r. Soc. Lond.* (In press.)

ROSSI, F. (1968). Bibliography on scanning electron microscopy. Published by Engis Equipment Company, Morton Grove, Illinois.

SMALL, E. B. and MARSZALEK, D. S. (1969). Scanning electron microscopy of fixed, frozen and dried protozoa. *Science, N.Y.* **163**, 1064–1065.

WELLS, O. C. (1969). Bibliography on the scanning electron microscope. *Proceedings 10th IEEE Ann. Symp. on Electron, Ion and Laser Beam Technology* (L. Malton, ed.). San Francisco Press.

WELLS, O. C. (1970). Bibliography on the scanning electron microscope. *Proceedings 3rd Annual Scanning Electron Microscopy Symposium*, pp. 509–524. IIT Research Institute, Chicago.

Author Index

A

Adams, I, 180, *208*
Allsopp, A., 213, *233*
Alvin, K. L., 223, *233*, 299, *306*
Amelunxen, F., 8, *16*
Anderson, T. F., 310, *315*
Andrew, R., 298, *306*
Armstrong, J., 49, 55, *65*

B

Backus, E. J., 288, *296*
Baker, E., 221, *234*
Baker, P. G., 49, *65*
Baldacci, E., 285, 291, *295*
Balme, B. E., *154*
Bancroft, N., 227, *235*
Banerjee, S., 246, *248*
Banerjee, U. C., 246, *248*
Barber, V. C., 308, *315*
Barnes, B., 181, *207*
Bauer, H., 288, *295*
Bé, A.W. H., 103, *111*, 181, 191, 205, *208*
Becker, B., 292, *295*
Beckett, A., 239, *249*
Berger, W. H., 201, *207*
Berggren, W. A., 181, *208*
Berkeley, M. J., 239, *248*
Biernat, G., 39, *65*
Bigelow, H. E., *282*
Bilal Ul Haq, U. Z., 180, *207*
Black, M., 180, 181, *207*
Bock, W. D., 135, *142*
Bolli, H. M., 188, 189, *207*
Boudreaux, H. B., 90, *93*
Boudreaux, J. E., 183, 188, *208*
Boulter, M. C., 213, 214, 227
Bowler, E., 221, *234*

Boyarskaya, R. V., 288, *296*
Boyde, A., 308, 310, 311, 313, *315*
Bradley, D. E., 288, *295*
Bradley, S. G., 288, *295*
Bramlette, M. N., 118, *121*, 180, 183, 191, *207*
Broome, C. E., 239, *248*
Brown, W. V., 227, *233*
Brunton, C. H., 48, *65*
Buchholz, J., 215, *233*
Bukry, D., 118, 120, *121*
Bund, C. F. von de, 90, *93*
Burrichter, E., 8, *16*
Bystricka, H., 183, *207*

C

Calonge, F. de D., 247, *249*
Čepek, P., 125, *142*
Cerceau-Larrival, M-T., 4, *16*
Chaloner, W. G., 214, *233*
Chandler, M. E. J., 214, *233*
Chenevière, E., 160, *176*
Cleve, P. T., 157, *176*
Cobden, R. H., 19, *36*
Coker, W. C., 271, 278, *282*
Colmer, A. A., 287, *295*
Cookson, I. C., 298, *306*
Couch, J. N., 271, 278, *282*
Couper, R. A., 148, *154*
Cross, T., 285, 291, *296*
Crowden, R. K., 3, *16*
Cullen, J., 10, *16*

D

Davies, F. L., 285, 288, 291, *296*
Davies, M. C., 288, *296*
Davis, P. H., 10, *16*

318

Author Index

De Beer, G. R., 29, *36*
Deflandre, 118
Demoulin, V., 271, 278, *283*
Desvaux, N. A., 278, *283*
De Toni, G. B., 157, *176*
Dietz, A., 288, *295*
Dilcher, D. L., 232, *233*, 297, 298, *306*
Dissing, H., *283*
Dusse, G., 90, *93*
Dring, D. M., *283*
Dronzek, B., 312, *315*

E

Echlin, P., *283*, 308, 312, 313, *315*
Eckblad, F. E., 239, *248*
Ellis, J. J., 239, 246, *248*
Emiliani, C., 191, *207*
Emiliani, C. E., 126, *142*
Evans, G. O., 69, *93*
Evko, E. I., 288, *296*

F

Fischer, E., *283*
Florin, R., 213, 214, 215, 222, 225, 227, *233*
Flugel, E., 188, *207*
Flynn, J. J., *284*
Franz, H. F., 188, *207*
Fraymouth, J., 240, 244, *249*
Funnell, B. M., 201, *207*
Furukawa, H., *284*

G

Geissler, U., 159, *176*
Gerloff, J., 159, *176*
Ghosh, B. K., 288, *295*
Giele, T., 8, *16*
Gilbert, E. J., 271, 278, *283*
Glauert, A. M., 288, *295*
Godwin, H., 298, *306*
Gooday, M. A., 247, *249*
Gran, H. H., 158, *176*
Grandjean, F., 78, 83, 84, 87, *93*
Gray, N., 215, *233*
Gray, T. R. G., 238, *248*

Greenhalgh, G. N., 247, *249*
Gregory, P. H., 268, *283*
Greguss, P., 212, *233*
Griffiths, D. A., 77, 78, *92*
Grunow, A., 156, *176*

H

Hamner, K. C., 221, *234*
Hanna, G. D., 128, *142*
Harborne, J. B., 3, *16*
Harland, W. B., 297, *306*
Harper, C. W., Jr., 55, *66*
Harris, T. M., 213, 221, 222, 227, *234*
Hatfield, H. L., 288, *296*
Hawker, L. E., 239, 240, 244, 245, 246, 247, *249*, *283*
Hay, W. W., 124, 125, 126, *142*, *143*, 181, 183, 188, 191, *208*
Hayes, T. L., 311, *315*
Heim, R., 252, 253, 259, 267, 268, *283*, *284*
Helle, W., 90, *93*
Helmcke, J-G., 159, *176*
Hendey, N. I., 158, 159, 173, 174, *176*
Hennig, W., 25, *36*
Heseltine, D. W., 239, 246, *248*
Heslop-Harrison, J., 309, *315*
Heslop-Harrison, Y., 214, *234*, 309, *315*
Heywood, V. H., 3, 10, 15, *16*, 308, *315*
Hinton, H. E., 18, 19, 23, 27, 32, *36*
Hollós, L., 239, *249*, *283*
Holt, S. C., 288, *295*
Honjo, S., 180, 181, *208*
Hopwood, D. H., 288, *295*
Horridge, G. A., 310, *315*
Hughes, A. M., 78, *93*, 94
Humphreys, W. J., 311, *315*
Hustedt, F., 158, 160, 174, *176*, *176*
Hutter, R., 291, *295*

I

Irving, E., 201, *208*

J

Jaanusson, V., 38, *65*
Jacot, A. P., 88, *93*

Johnson, C., 227, *233*
Johnson, V., 308, *315*
Jones, D., *283*
Jongbeloed, W. L., 103, *111*
Jope, M., 39, 55, *65*
Juniper, B. E., 221, *234*

K

Kawato, M., 287, *295*
Kilpper, K., 214, *234*
Konev, Y. Y., 292, *296*
Krasilnikov, N. A., 292, *296*
Krieger, W., 159, *176*
Kudrina, E. S., 288, *296*
Küster, E., 288, *296*
Kützing, F. T., 156, *176*

L

Lampen, J. O., 288, *295*
Lance, G. N., 10, *16*
Lange, M., 239, *249*, *283*
Laubenfels, D. J. de, 215, *234*
Leadbetter, E. R., 288, *295*
Lechevalièr, H. A., 292, *295*
Lechevalièr, M. P., 292, *295*
Lee, B., 219, *234*
Litke, R., 212, *234*
Lloyd, C. G., 268, *283*
Locci, R., 285, 291, *295*
Loeblich, A. R., Jr., 135, *142*
Long, W. H., *283*
Lorch, J., 223, *234*
Lukivanovich, V. M., 288, *296*

M

McCoy, E., 287, *295*
Macfarlane, D., 69, *93*
McIntyre, A., 103, *111*, 181, 191, 205, *208*
McNeill, J., 3, *16*
Mädler, K., 214, *234*
Mägdefrau, K., 217, 223, *234*
Maksimov'a, T. S., 288, *296*
Malençon, G., 267, 274, *283*
Mann, A., 158, *177*
Manton, S. M., 92, *93*

Märkel, K., 88, *93*
Marszalek, D. S., 126, *143*, 308, *315*
Martini, E., 118, *121*, 191, 205, *208*
Martinsson, A., 107, *111*
Mathews, J., 288, *295*
Matthews, H. I., 246, *249*
Mayfield, C. I., 288, 293, *296*
Mégnin, P., 70, 71, *94*
Meyer, I., 88, *93*
Milne, A., 32, *36*
Minbaen, R., 292, *296*
Mohler, H. P., 183, 188, 191, *208*
Moor, H., 288, *296*, 310, *315*
Morgen Roth, K., 8, *16*
Morosov, V. M., 292, *296*
Muhlethaler, K., 288, *296*
Muir, M. D., 299, *306*

N

Nachiò, M., 180, *208*
Nautiyal, D. D., 212, 227, *234*
Nixon, H. L., 268, *283*
Nixon, W. C., 181, *208*
Nöel, D., 1, 115, 118, 120, *121*, 180, 188, *208*

O

Oatley, C. W., 3, *16*, 181, *208*
Okada, H., 180, 181, *208*
Oudemans, A. C., 69, *94*
Owen, G., 48, *66*

P

Paden, D., 312, *315*
Palmer, J. T., *283*
Panti, D. D., 212, 226, 227, *234*
Pantin, C. F. A., 1, 3, *16*
Parke, M., 180, *208*
Parker, P. F., 3, *16*
Pearce, R. F. W., 181, *208*
Pease, R. F. W., 311, *315*
Pegler, D. N., *284*
Peragallo, H., 158, *177*
Peragallo, M., 158, *177*

Perch Nielsen, K., 118, *121*, 183, *208*
Perreau, J., 252, 253, 268, *284*
Peterson, M. N. A., 201, *208*
Phleger, F. B., 126, *143*
Picksak, T., 8, *16*
Pilát, A., 271, 278, *284*
Plunkett, O. A., *283*
Porsch, O., 231, *234*
Preikstas, R., 181, *208*
Preobrazhenskaya, T. P., 288, *296*
Priestley, J. H., 219, *234*
Pritchard, A., 156, *177*

R

Ratcliffe, A., 78, *94*
Reimann, B., 159, *176*
Reinhardt, P., 116, 118, *121*
Remsen, C. C., 288, *295*
Reyre, Y., 8, *16*, 145, 146, 148, 150, 152, *154*
Riedel, W. R., 180, 183, 191, *207*
Ritzi, D., 288, *295*
Robertson, P. L., 78, *94*
Robinson, D. M., 90, *94*
Robinson, T., 211, *234*
Rolland, L., 271, 278, *284*
Roper, F. C. S., 156, *177*
Ross, R., 155, 176, *177*
Rossi, F., 308, *315*
Roth, K., 8, *16*
Roth, P. H., 183, 188, *208*
Rowley, J. R., 282, *284*
Ruddick, S. M., 293, *296*
Rudolf, H., 217, 223, *234*
Ruffle, L., 212, *234*

S

Sandberg, P. A., 124, *142*, *143*, 181, *208*
Saxena, S. K., 246, *248*
Schmidt, R. R., 183, 188, *208*
Schütt, F. 157, *177*
Schwendener, S., 230, *234*
Scott, F. M., 221, *234*
Sharples, G. P., 289, 293, *296*

Sheals, J. G., 69, 84, *93*, *94*
Shinobu, R., 287, *295*
Siegel, S., 182, *208*
Sims, P. A., 176, *177*
Small, E. B., 308, *315*
Smith, A. G., 201, *207*
Smith, H. L., 156, *177*
Srivistava, G. K., 227, *234*
Stace, C. A., 213, *234*
Subramanian, C. V., 247, *249*
Sullivan, F. R., 183, 191, *207*
Suzuki, Y., 298, *306*
Sveshinkova, M. A., 288, *296*
Swinburne, T. R., 246, *249*
Sylvester-Bradley, P. C., 99, *111*
Szafer, W., 214, 217, *235*

T

Tappan, H., 135, *142*
Tamm, S. L., 310, *315*
Thomas, B., 239, *249*
Thomas, H. H., 227, *235*
Torre, M. de la, 240, 244, *249*
Towe, K. M., 55, *66*
Townrow, J. A., 215, 223, *235*
Tregner, H. D., 288, *296*
Triebel, E., 131, *143*
Tsi-Shen, J., 292, *296*
Tsyganov, V. A., 292, *296*

U

Uozumi, T., *284*

V

Vahl, J., 8, *16*
Van Heurck, H., 157, *177*
Vittadini, C., 239, *249*

W

Wartmann, R. von, 221, *235*
Watabe, N., 120, *121*
Watson, L., 10, *16*
Watson, R. W., 213, *235*
Wayte, R., 312, *315*

Wells, O. C., 308, *315*
Weyland, H., 212, *235*
Wildermuth, H., 288, *296*
Willetts, H. J., 238, 247, *249*
Williams, A., 39, 40, 48, 49, 55, 58, 59, 60, *65, 66*
Williams, S. T., 285, 288, 289, 291, 292, 293, *296*
Williams, W. T., 10, *16*
Wood, C., 308, 310, 311, 313, *315*
Wright, A. D., 55, 59, 60, *66*
Wright, R. C., 126, *143*

Yamaguchi, T., 292, *296*
Young, T. W. K., 239, *249*, *284*

Z

Zachvatkin, A. A., 78, *94*
Zeller, S. M., *284*
Ziegenspeck, H., 230, *235*

Subject Index

Numbers in *italic* refer to Plates illustrating the material described

A

Abies, 217, *218*, 221, 222, *224*, 230, 231
  alba, 217, 221, 227, *228*
Absidia, 239
Acanalonia, 21
Acanalonidae, 21
Acanthocrania, 59
Acari, 67–93
Acarines, 67–93
Acarus, 77, 79
  "*casei*", 68
  chaetoxysilos, 75
  domesticus, 68
  farris, 74, 78, *82*
  gracilis, 79
  immobilis, 78, *82*
  macrocoryne, 75
  nidicolous, 73
  siro, 68, 70, 71, 78, 79
  tyrophagoides, 79
Acrotretida, 39, 55, 61
Actinobifida, 292
Actinochaeta, 69, 71, 78, 92
Actinocyclus, 156, 173
Actinomycetales, 285
Actinomycetes, 285–294
Actinoplanes, 290
Actinopycnidium, 292
Actinosporangium, 292
Aeropyles, 17, 21, *28*, *30*, 35
Agaricales, 268
Agaricées, 252
Agrotidae, *28*
Ahmuellerella, 116
  octoradiata, 116, *117*
Algae,
  chrysophitid, 180

coccolith-bearing, 179–207
  pelagic, 180
Allium cepa, 221
Amphipentas, 156, 157, 166, 175
Amphitetras, 156, 157, 164, 175
Anactinochaeta, 69, 92
Anaulus, 176
Anguliferae, 156
Anopheles, 21
Antheraea pernyi, *28*
Apterocladus, 147
Araucaria, 227
  australis, 154
  bigwelli, 148
  brasiliensis, 148, 153
  pollenites, *149*, *151*, 153
Araucariacées, 148, 153, 154
Araucariacites, 147, 153, 154
Arctia caja, *28*
Arctiidae, *28*
Argynnis selene, *28*
Artemia, 311
Arthropods, 293
  chelicerate, 67–93
Articulata, 38, 40–45, 61–65
Ascomycetes, 297, 298
Ascospores, 239, 244
Aspergillus, 246
Astérosporales, 252
Astigmata, 69, 71, *74*, 75–81
Aulacodiscus, 176
Auliscus, 158, 176
Aureotactae, 259
Aurila, 102
  convexa, *98*, *107*

B

Bacillus, 292

Bacteria, 285
Basidiospores, 239, 246, 251–282
Bennettitales, 227
Bensonocythere whitei, 110
Biddulphia, 155–159, 172, 173, 175
  armata, 173
  aurita, 174
  biddulphiana, 159, 161, *163*, 172
  mobiliensis, 173
  parallela, 174, 175
  polyacanthos, 174, 175
  reticulata, 168, *169*, 170, 174
  rhombus, 162, *167*, 170, 173
  toumeyi, 159, 172
  zanzibarica, 174, 175
Biddulphiaceae, 155–176
Biddulphioideae, 176
Billingsella, 65
  lindstromi, *52*, 54, 64
Billingsellacea, 54, 62–64
Billingsellidae, 64
Bimuria, 54, 63
  buttsi, 52
Bobinella, 49, 62
  kulumbensis, 50
Boletales, 268
Bone, 307
Boraginaceae, *13*
Bovista plumbea, 268, 269, *281*
Braarudosphaera bigelowi, *195*, 206
Braarudosphaeridae, 191, *195*, 202, 204, 205
Brachiopoda, 37–65
Brassica, 4
Broinsonia cribella, *187*, 190, *193*
Bryobia, 87
Bythoceratina scaberrima, 101, *106*

C

Cadomellacea, 39
Callimothallus, 303
  pertusus, 305
Caloglyphus berlesei, 80
  mycophagus, 82
  redikorzevi, 76

Calvatia, 270
  candida, 270, *281*
  lilacina, 270, 272, *281*
Calvolia, 74
Carabidae, 27
Caractères, *see* Characters
Carpoglyphus lactis, 74, 76, 82
Carposphere, 15
Caucalis, 4, 5, 12
  platycarpos, 9
Cerataulus, 156–158, 164, 174, 176
  turgidus, 164, *169*, 173
Chaetaxy, 70–73, 83, 86, 92
Chaetosciadium, 6, *11*
  trichospermum, *11*
Champignons, 251–282
Characters, taxonomic, 4, 8–15, 31, 75, 77, 78, 81, 87, 90, 92, 103, 107, 115, 116, 121, 146, 153, 161, 162, 213–223, 231, 232, 239, 240, 245, 255, 257, 266, 272, 278, 285, 292
  unit, 14
Chelicerates, *see under* Arthropods
Chiphragmalithus quadratus, 189, *193*, 206
Chlamydospores, 246
Chorion, *see* Insects, eggs
Chrysopa, 21, *24*
Chytrids, 297
Classification, 8, 10, 12, 19, 23–27, 35, 37, 69, 71, 75, 83–86, 92, 93, 114, 126, 146, 155–176, 181, 237–248, 265, 285, 291, 297, 304. *See also under* Pollen
Classopollis, 152
Coccolithites cribellum, *187*
Coccolithophoridées, 113–121
Coccoliths, 15, 113–121, 129, 179–207
Coccolithus huxleyi, 120, 205
Coelopidae, 23
Computer processing, 8, 141
Conidia, 239, 246, 247
Conifers, *see* Gymnosperms
Conninghamella, 239, *243*
Coprophila, 23
Cordaites, 221
Cordiluridae, 22, 23
Coscinodiscus, 173

Crania, 58–60
  anomala, 45, 55, *56*
Craniacea, 55, 59–61, 64, 65
Craniops implicata, *56*
Craniopsidae, 60, 61, 64, 65
Cretarhabdus, 118
Cruciferae, 4
Cruciloculina triangularis, *138*
Cryptomeria, 217, 227
  anglica, *303*, 304
  japonica, 217, 226
  rhenana, 217, *220*, 225, 226
Cryptostigmata, 69, 71, 81–91
Culex pipiens fatigans, 26
Culicidae, *26*
Cupressacées, 153, 278
Cupressacites, *149*, *151*, 152
  oxycedroides, *149*, *151*, 153
Cuticle, 5, 6, 211–233
Cuticular flange, 222
Cylapinae, 21
Cylindralithus biarcus, *119*, 120

D

Daucus, 4
Davidsoniacea, 64
Denticella, 156, 157
Dermanyssus gallinae, *89*, 91
Dermatophagoides farinae, 76
Deuteromycetes, 246, 297
Dianthus plumarius, 309
Diatoms, *195*, 307
Dictyocephalus curvatus, 276
Dictyonellidina, 60, 61
Diptera, 21, 23, 25
  Cyclorrhapha, 19
Discoaster, 183, 190, 191, 201, 204
  barbadiensis, 183, *185*, 188, 189, 200–204, 206
  elegans, 183
  lodoensis, *185*, *193*, *195*, 204, 206
  multiradiatus, 183, *185*, 189, 204–206
  multiradiatus robustus, 183, *195*
  ornatus, 187–189, 206
  sublodoensis, *185*, 189, *195*, 204, 206

Discoasters, 180, *195*, 202, 205
Discopodinae, 252, 253, 256, 259
Drosophila, 18, 23
  flavicola, 23
  gibberosa, 22
  melanogaster, 18
  sexvittata, 23
Drosophilidae, 22, 23
Dryomyza, 23
Dryomyzidae, 23

E

Echium callithyrsum, *13*
Eggs, insect *see under* Insect
Eiffelithus, 116
  eximius, 117
Elaphomyces, 240
  asperulus, 239, 240
  granulatus, 239, 240, *243*, 244
Elodea, 309
Empididae, *20*
Enteletacea, 54, 62
Epidermis, leaf, 6, 212–233
Episyrphus balteatus, *20*
Epithelium of Brachiopod shell, 38–41, 55, 58
Ericsonia, *195*
  alternans, 190, *193*
  ovalis, 190, 191, *193*, *195*, 206
Eucyrtidae, 23
Euglena, 313
Eumusca, 23
  autumnalis, 22
Eunotogramma, 175
Euodia, 156
Eupodiscaceae, 157, 158, 176
Eupodiscus, 176
Euphthiracaridae, 87, 88
Euphthiracarus, 88, 90
  cribrarius, 89
Euptyctima, 84, 86, 87
Evolution, 17–37, 60, 62, 63, 65, 124–126, 233, 252, 265

F

Famulus, 81, *82*

Fannia, 35
  armata, 24
  atripes, 24
  canicularis, 24
  coracina, 24
  nidica, 24
Foraminifera, 136–140
  benthonic, 128, 191
  pelagic, 128
  planktonic, 103, 126–129, 188–191, 201
Forcellinia wasmanni, 72
Fossils, 95–110, 123–141
  angiosperm, 212
  Biddulphiaceae, 164, 170, 174–176
  brachiopod, 37–65
  coccolith, 113–121
  conifer, 211–233, 298
  epiphyllous fungus, 297–306
  nannoplankton, 179–207
Frenelopsis, 298, *300*
Fruits, 12
  of Umbelliferae, 1–15
Fungi, 237–248, 285–294, 251–282, 297–306

G

Gartnerago, 120
  obliquus, *119*
Gasteromycetes, 247, 267–282
Geasters, 268
Geastrum, 270, 274
  asper, *277*
  caespitosum, 270, *275*
  hariotii, 270, *275*
  mirabile, 268
  pectinatum, 270, 272, 274, *275*
  rufescens, 270
  saccatum, 270, *275*
  subiculosum, 268, *275*
  velutinum, 269, 270, *275*
Geholaspis, 91
  longispinosus, 89
Genea, 245
Geometridae, *28*
Geopora, 246
Globorotalia cultrata, *130*, *132*, *134*, *137*, 141

Glossopteris, 226
Glycyphagus destructor, *80*, 81
  domesticus, *80*, 81
Gohieria fusca, 76
Grandjean's organ, 74, 75
Gymnosperms, 145–154, 211–233, *see also* Fossil conifers

H

Hebecnema, 23
  umbratica, 22
Heliochromae, 253, 259
Helicosphaera, 206
  seminulum, *193*, 206
Helicosphaeraceae, *193*, 204, 205
Hemiaulioideae, 176
Hemiaulus, 176
Hemiptera, 21
Hemithirispsittacea, 42
Heterocampa manteo, 30
Heterophyllae, 252, 264
Hirtodrosophila, 23
  sexvittata, 23
Horn, respiratory, *see* Respiratory horn
Hydnoterya, 245
Hydrosera, 156, 176
Hymenogaster, 246
  hessei, *243*
Hyménomycètes, 266
Hymenoptera, 23
Hypodermis, 223, 230
Hysterangium, 239, 246
  separabile, 246
Hysteriaceae, *300*, 301

I

Ichthyura albosigma, 30
Idiostatus aequalis, 26
  siskiyou, 26
Ilyocypris nitida, 110
Inaperturopollenites, 147, 153
  cf. *Araucariacites australis*, *149*, *151*, 154
  cf. *Araucariapollenites*, *149*, *151*
  giganteus, *149*, *151*, 153
Inarticulata, 38, 55–60

Indumentum, 6, *13*
Information, 2, 8–12, 95, 105–110, 123–142
Insects, 67
  eggs, 5, 17–36
Isthmia, 156
Isthmolithus recurvus, 187–189, 191, *193*, *195*, 199, 206

K

Kamptnerius, 120
  magnificus, *119*
Koninckina leonhardi, *50*
Koninckinacea, 39, 54, 61, 62
Kotujella, 54, 62, 64
Kutorginida, 61, 64

L

Labiatae, *13*
Lacazella mediterranea, *50*
Lactaires, 252, 253
Laffittius, 120
Lagarosiphon major, 312
Laricoidites, 147
Leioaletes, 147
Lepidoptera, 21
Leptocera, 23
Leudugeria, 175
Lignin, 226–231
Limnophora, 23
Lingulida, 60, 61
Lophodolithus nascens, *193*, 206
Lycoperdaceae, 267, 278
Lycoperdon atropurpureum, 272, 274
  echinatum, 268, 269, 272, *281*
  pyriforme, 268
  umbrinum, 268, 272, *281*

M

Magasella sanguinea, *46*
Markalius circumradiatus, 118, *119*
Marthasterites contortus, 187–189, 206
  tribrachiatus, 183, *185*, *187*–190, 206
Massilina protea, 138
Masulostrobus warrenii, 147
Mecoptera, 25

Mericarp, 1–15
Mesostigmata, *89*, 91, 92
Metasequoia glyptostroboides, 217, *218*, 221, 222, *224*
Methods, *see* Techniques
Micrantholithus attenuatus, 206
Micro-algae, 312, 313
Microbes, 307
Microbispora, 290, 292
Microechinospora, 292
Microellobosporia, 292
Micromonospora, *290*
Micropeltaceae, 299
Micropolyspora, *290*
Microscope, light, 5, 8, 17, 71, 81, 83, 86–93, 121, 127–131, 138, 146, 147, 152–154, 181–183, 212–217, 222–232, 238, 240, 248–292, 299–309
  projection X-ray, 103, 105, 107
  transmission electron, 1, 2, 39, 40, 91, 103, 105, 114, 115, 129, 131, 159, 180–183, 188, 230, 237, 240, 246, 247, 286–293, 305–311
Microthallites, 302, *303*, 304, 306
Microthyriales, 298, 299, 305
Microstrobus sommervillae, 223
Microtritia, 88
  minima, 90
Miridae, 21
Mites, 67–93
Moorellina, 49
Mucor, 238
  mucedo, *243*
Mucorales, 238, 239
Musca autumnalis, 22
  sorbens, 19, *26*
  vetustissima, 19
Muscidae, *20–26*, 35
Muscinae, 19
Mutationella, 49
  podolica, *46*
Mycenastrum, 280
  corium, 271, 278–280
  chilense, *279*, 280
  phaeotrichum, *279*, 280
  spinulosum, 280

Mydea, 23
Myospila, 23
Myriostoma, 274
    coliforme, 269, 274, 277

N

Nadata gibbosa, 30
Nanofacies, 15
Neococcolithes dubius, *193*, 206
Nepa rubra, 22
Nepidae, 21–23
Neuroptera, *24*
Nigricantinae, 252
Nilssonia tenuicaulis, 222
    tenuinervis, 222
Nisusia, 54, 62, 64
    ferganensis, *52*
Nothoceratium, 157
Notodonta dromedarius, 30
Notodontidae, *30*
Notogaster, 84, 88
Notosaria, 41, 48
    nigricans, 40–42, 45
Numerical taxonomy, *see* Taxonomy
Nymphalidae, *28*

O

Obolellida, 60, 61, 64, 65
Ochripleura plectra, 28
Odontella, 156–158, 174–176
Opalina, 310
Orbulina universa, 128
Oribotritiidae, 87
Orlaya, 4
Orthacea, 54, 62
Orthellia caesarion, 26
Orthida, 54, 55, 60, 62, 63
Orthophlebiidae, 87
Orusia, 54
Orygma, 23
Ostracods, 128, 136
Ozaenini, 27

P

Palynology, *see* Pollen

Papillate, 222
Paramecium, 313
Pattrayella, 176
Paussidae, 27
Pelliculariae, 252, 256, 258, 266
Pemma papillatum, *195*, 206
Pentamerella cf. *lingua*, *50*
Pentamerida, 49, 60, 62
Pergamasus, 91
Periostracum, 38–49, 59, 62–64
Petrocrania, 59
Peziza proteana var. *sparassoides*, 246
Pezizales, 244
Phellorina delestrei, 269, *273*, 276
    inquinans, 276
    leptoderma, *273*, 276
Phellorinés, 274
Phellorinaceae, 267
Pheosia gnoma, 30
    tremula, 30
Phleridosa flavicola, 23
Phragmatobia fuliginosa, 28
Phthacaridae, 84, 87
Phycomyces, 239
Phyllosphere, 15
Phylogeny, 17–36, 39, 60, 61, 71, 114, 121, 181, 223, 252
Picea, 230
    abies, 217, 227, *228*, 231
Pinguicula grandiflora, 214
Pinus, 213, 215, 217, *220*–224, 230, 231
    peuce, 217, *218*, 222, 223, 227, *228*, 230
    sylvestris, 217, 227
Plasmalemma, 41, 44
Plastron, 17–*24*, *26*, *28*, *30*, 35
Platycheirus peltatus, 20
Platypalpus pallidiventris, 20
Plectambonitacea, 54, 60, 61, 63, 64
Podocarpacées, 147
Podocarpus, 215, 232
    setiger, 223
    Sect. *Stachycarpus*, 232
Podorhabdaceae, 118
Podozamites, 147
Pollen, 5, 8, 12, 96, 145–154, 298, 307
    classification of, 147–154

Polietes lardarius, 20
Polypodorhabdus, 118
Pontosphaera, 206
　fimbriata, 206
　ocellata, 206
　pectinata, 206
　plana, 206
　pulcher, 206
　rimosa, 206
Pontosphaeraceae, 191, *195*, 202, 204, 205
Porambonitaceans, 49, 62
Porpeia, 156, 176
Productaceans, 54
Productids, *52*, 55
Prostigmata, 87, *89*, 90
Protozoa, 307, 312, 313
Protozyga, 49
　rotunda, 46
Pseudauliscus, 176
Pseudacrania, 59
Pseudogomphus, 266
Psophosphaera, 147
Publication, problems of, 105, 107, *see also* Information
Pucciniales, 297
Pyrgo comata, 138
　denticulata, 138
　elongata, 138
　fomasinii, 138
　murrhina, 138
　subsphaerica, 138

Q

Quench freezing, 312–314
Queletia, 270
　mirabilis, 265, 270
Quinqueloculina, 138
　tricarinata, 138

R

Radicantes, 252, 258, 259
Ranatra fusca, 22
Resolution, 2, 39, 69, 96, 100, 103, *104*, 127, 129, *132*, 181, 217, 291, 299, 307
Respiratory horns, 18–23

Rhabdosphaera, 191, 202, 204–206
　creba, 206
　perlonga, *195*, 206
　truncata, 206
Rhingia campestris, 20
Rhipidomella, 50
Rhizoglyphus, 75
Rhizopus, 239
　sexualis, 238, *241*, 247
Rhizosphere, 15
Rhynchonellida, 40–49, 54, 55, 62
Rhynchospora, 48
Rhysotritia, 88, 90
　ardua, 88, 90
　duplicata, 88, *89*, 90
　minima, 88
Root hairs, 312
Rostricellula, 48
　lapworthi, *42*
Russula, 252–267
　acutispora, 256, *258*, 261, 266
　africana, *258–261*
　alveolata, 259, *262*, 263
　annulata, 253, 267
　　var. *molochina*, 253, *255*, 258, 266
　　subsp. *parasitica*, 253, *255*, 259, 266, 267
　　f. *violacea*, 253
　carbonaria, 252, 255, 267
　carmesina, 253, 256, *261*
　cyanea, 253–258,
　cyanoxantha, 265
　　var. *variata*, 265
　discopus, 256, 258, *261*,
　echinosperma, 259, *262*, 263 266
　mimetica, 259, *262*, 263, 267
　molochina, 266
　papillata, *262*, 263, 266
　radicans, 267
　viridicens, *262*, 263, *264*, 266

S

Saldidae, 19
Sampling, 10
Saturniidae, *28*
Schuchertella haraganensis, *52*

Sciadopitys, 231
  verticillata, 217, *218*, 227
Sclérodermes, 278
Sclerotinia fructigena, 247
Scopeuma stercorarium, 22
Seeds, 12, 307
Semiothisa signaria dispunctata, 28
Sepsidae, 22, 23
Sepsis violacea, 22
Sequoia couttsiae, 217, *218*, *220*, 222, 225, 226
Sequoiadendron giganteum, 217, *220*, 225–228, 230
Shell, ostracod, 95–110, see also Skeleton
Sideritis canariensis, 13
Sinapis alba, 312
Siphonaptera, 25
Skeleton, brachiopod, 37–65
Sleiatroglodytophila, 97
Smilax, 302, *303*
Solenidia, 78, 81–87
Sowerbyella variabilis, *52*
Sphaeroceridae, 23
Sphaeropollenites, 147, 148
  scabratus, *149*, *151*, 153
Spiriferida, 49, 54, 62, 63
Spiriferna, 49
  walcotti, *46*
Sphenolithus furcatolithoides, *187*–191, *193*, *195*, 198, 200, 201, 206
Spores, 15, 237–247, 251–282, 285–293
Steganacarus, 84–87
  magnus, 86, 87
  striculus, *85*–87
Stereo-pairs, 100, 103, 105, *106*
Strictodiscus, 157, 158, 173
Stomata, 214, 215, 223, 230, 231
Stomiopeltites, 301, 304
  cretacea, 299, *300*
Stomiopeltoideae, 299
Stradneria, 118
  limbicrassa, *117*
Stratigraphy, 125, 126, 183, 188, 189
Streptomyces, 286, 288, 289, *292*–*294*
  violaceoruber, 288
Strophomenacea, 64

Strophomenida, 49, 54, 55, 60–63, 65
Suidasia, 74
Supra-coxal seta, 78, 80
Syrphidae, *20*, 35
Syrphidius ribsii, 20

T

Taxacites, 152
  sahariensis, *149*, *151*, 152
Taxodiaceae, 222, 227, 230
Taxonomic characters, see Characters
Taxonomy, multivariate, 3
  numerical, 14, 31, 84, 86
Taxus, 213
Techniques, 39, 40, 95–105, 159, 181, 182, 213, 214, 237, 238, 248, 287–291, 307–314
Telletia caries, 246
Terebratulida, 40–49, 54, 59, 61, 63
Terebratulina retusa, *46*
Termatophylidea, 21
Tetranychidae, 90
Tetranychus, 87, 90
  cinnabarinus, *89*–91
  urticae, 87, *89*–91
    f. *dianthica*, 91
Tettigometra, 21
Tettigometriidae, 21
Thecideidina, 49, 61, 63
Thecidellina, 49
  barretti, 45
Thermoactinomyces, 292
Terpsinoe, 175
Terpsinoeae, 156
Torilis, 5–7, 12
Tradescantia, 313
  bracteata, 312
Transmission electron microscope, see under Microscope
Trematobolus, 60
  pristinus bicostatus, *56*
Tribolium confusum, 311
Triceratium, 155–158, 161, 174–176
  antediluvianum, 164, 168, 171, 174, 175
  arcticum, 173
  castelliferum, 172

Subject Index

Triceratium (contd)
  crenulatum, 172
  dubium, 174
  flavus, 170, *171*, 174, 175
  flos, 172
  formosum, 161, *165*, 173
  fractum, 173
  glanduliferum, 172
  inelegans, 173
  majus, 172
  montereyi, 173
  morlandii, 172
  nova-zealandicum, 172
  pentacrinus, 166, 168–170, 174
  planoconcavum, 173
  polycistinorum, 172
  reticulum, 173
  rugosum, 173
  secedens, 174
  shadboltianum, 173
  spinosum, 159, 161, 162, 164, *167*, 170, 173
  stokesianum, 159, 161, *163*, 172
Trichaster melanocephalus, 269, 270, 274, *277*
Trigonium, 157, 158, 161, 173, 175
  arcticum, 161
  margaritaceum, 173
Trimerellacea, 38
Trinacria, 176
Triplesiidina, 39, 54, 55, 63
Tritaenia, 217
Tropiduchidae, 21
Tsuga, 147
  heterophylla, 217, 219, 223, *224*, 226, 230
Tsugaepollenites, 147
  mesozoicus, 147
Tuber, 244
  aestivum, 245
  puberulum, 245
  rufum, 245
Tuberales, 244–246

Tulostoma, 265
  barlae, 265, 269, 270, 272
  berkeleyii, 265
  brumale, 272
  campestre, 268, 272
  chevalieri, 265
  floridanum, 265, 270
  volvulatum, 272
Tulostomataceae, 267, 270
Tulostomes, 268
Tyroglyphus siro, 70
Tyrolichus casei, 74
Tyrophagus, 80
  nieswanderi, 76, 82
  peruiciosus, 80
  putrescentiae, 74

U

Umbelliferae-Caucalidae, 1–15
Ustilaginales, 246

V

Valve, brachiopod, *see* Skeleton of Biddulphiaceae, 155–176

W

Waltonia inconspicua, 45, *46*
Whetstonia strobiliformis, *273*, 276

X

Xylariaceae, 247

Z

Zea mays, 312
Zonalapollenites, 152, 154
  cf. *segmentatus*, *149*, *151*, 154
Zonation, *see* Stratigraphy
Zygoceros, 156, 162, 174, 176
Zygodiscus, 206
  sigmoides, 206
Zygolithaceae, 115, *193*, 204, 205
Zygorhynchus moelleri, 238, 241